Wetland Weeds

Causes, Cures and Compromises

Nick Romanowski

PUBLISHING

National Library of Australia Cataloguing-in-Publication entry

Romanowski, Nick, 1954–

Wetland weeds : causes, cures and compromises/By Nick Romanowski.

9780643103955 (pbk.)
9780643103962 (epdf)
9780643103979 (epub)

Includes bibliographical references and index.

Aquatic weeds.
Invasive plants.

581.76

Published by

CSIRO PUBLISHING
36 Gardiner Road, Clayton VIC 3168
Private Bag 10, Clayton South VIC 3169
Australia

Telephone: [+613] 9545 8555
Local call: 1300 788 000 (Australia only)
Fax: +61 3 9662 7555
Email: csiropublishing@csiro.au
Web site: www.publishing.csiro.au

Front cover: Arum lily

All photographs are by the author unless otherwise noted

Set in Adobe Minion 11/13.5 and Adobe Helvetica Neue LT
Edited by Anne Findlay, Editing Works Pty Ltd
Cover design by Samantha Duque
Text design by James Kelly
Typeset by Desktop Concepts Pty Ltd, Melbourne
Printed by Ingram Lightning Source

Feb26_RP_ILS

Contents

Introduction

I began experimenting with indigenous wetland and aquatic plants in the 1970s, mainly looking at their diverse roles as habitat for animals. My primary interests and qualifications were in zoology, with a particular focus on fishes, aquatic invertebrates and some frogs, though wetland plants had long intrigued me in their own right. Helen Aston's *Aquatic Plants of Australia* was published while I was still a zoology student, and turned out to be the tipping point for my latent interest in the plant-related aspects of wetlands.

By the early 1980s I was experimenting with commercial production of indigenous wetland plants, with the idea that planned introduction of plants and animals into created wetlands could help to replace the ever-growing number of drained and damaged wetlands, but the steep learning curve I was going through included playing around with diverse exotic species. When my nursery finally opened, our early catalogues included a range of exotics on the advice of water garden experts who had persuaded me that no-one could possibly make a living from indigenous species alone (they were wrong). Within several years the catalogue introduced a section warning of the weed potential of many non-native plants, and this snowballed into an ever-expanding list of plants we were no longer prepared to sell, even though they were still available elsewhere. It was a novel approach to selling exotic aquatics which doesn't seem to have taken the world by storm, for obvious reasons.

Eventually, the nursery became what I had originally intended it to be, specialising in habitat enhancement and water treatment in farm dams, grey-water systems and the like, but we are still haunted by ghosts from our earliest years. Many of my books have been reviewed in the journal of the International Water Lily Society, and as our business name wasn't registered overseas it was later snapped up by a water garden nursery in the USA. This would not be a problem except that they also claimed the domain name, so readers looking for a *Dragonfly Aquatics* website will be surprised to find nothing on it but exotic weeds! The only way to contact the original, Australian *Dragonfly Aquatics* (which does not maintain a website) is by phone or email.

Despite such irritations, in retrospect my hands-on experience with most of the introduced weeds and potential weeds discussed in this book has turned out to be useful. Even within the well-defined precincts of the nursery it was difficult to eradicate some of the plants we decided should not be sold, and in the course of writing and research it was disturbing to learn how many of these are still legally available. It is time to pull the plug on further imports of all potential weeds, terrestrial or aquatic, regardless of their origins, supposed useful properties, or ornamental value. Wetland weeds make up just a small proportion of our feral flora but they are a very real threat to agriculture, industry, such luxuries as a daily shower, and even the quality of the water we drink.

Acknowledgements

In the course of writing this book, many people have helped, advised, sent me plants for photography or directed me to photogenic locations, located books and other materials, and even disagreed with me in constructive ways. I would like to thank Paul Adam, Craig Allen, Helen Aston, Kate Blood, Kathy Bown, John Dodson, Tim Entwisle, Nola and Michael Fenech, Lalith Gunasekura, Pierre Horwitz, Surrey Jacobs, Andrew Paget, Jessica Seidel, Kelly Snell, Val Stajsic, Neville Walsh, Darren Wilkie, Dave Wilson and Karen Wilson for their various and diverse inputs over the years. Although nearly all of the photography in this book is my own, I am grateful to Stuart Willsher for permission to use his photo of *Spartina anglica*. Once again, I am grateful to Ted Hamilton at CSIRO Publishing for pushing me into another project. Without his encouragement and patience, it might have just gathered dust for many years.

1
What is a weed?

The simplest definition of a weed is 'a plant out of place', so even an indigenous species with significant habitat values in natural wetlands may be perceived as a serious weed in a rice field 100 metres away, or if it is introduced into another part of the country where it does not belong. Some plants may be weedy because they form monospecific (single-species) stands by killing off competing plants with a chemical arsenal produced for that purpose. Others only become a problem when growing in unnaturally nutrient-rich waters or soils, and raise no concerns in less polluted habitats.

However, the great majority of weeds are introduced plants which have run wild in new habitats, where they are free of the pests, diseases and competing species that act as checks and balances on their growth in the environments where they originally evolved. Introduced weeds are now a universal phenomenon, invading and altering wild habitats, competing with crops, poisoning animals and blocking drainage canals. Many of them were introduced as agricultural plants with no thought that improvements in pasture production would be more than offset by the cost of repeatedly clearing them from irrigation channels, crops and the wider environment.

An estimated two-thirds of all terrestrial weeds come from the garden trade, and three-quarters of the established wetland weeds in Australia were introduced as ornamentals for water gardens and aquariums. Aquatic and wetland plants may only be a small part of the weed problem but they have a critical impact on the quality of fresh water, a critical resource for the economy essential for industry, agriculture, and even the luxury of a daily shower. This book considers only those weeds that grow in water, or along the wet to waterlogged fringes of wetlands and streams. On this driest of livable continents, they dramatically increase water losses through evaporation, clog streams and channels, and contaminate drinking water with unpleasant and sometimes toxic wastes. Many of them also invade natural wetlands, already suffering from drainage, invading animals, pollution and general neglect.

While writing this book I found it disturbing to find how easily I could locate weeds I had not encountered before, often hundreds of kilometres closer to home than they had been a decade ago. Japanese kelp (*Undaria pinnatifida*; see plate 32) had been 'conveniently' introduced on yachts into the nearby harbour at Apollo Bay. New garden escapes such as tall willowherb (*Epilobium hirsutum*; see plate 29) sprang up along the Barwon River in Geelong, and it only took a few minutes to locate dense stands of alligator weed (*Alternanthera philoxeroides*; see plate 1) on a trip to visit family in Melbourne.

Such weeds are likely to become still more far-ranging as the effects of global warming become more pronounced, spreading further into areas with an increasingly suitable climate for them. This is particularly true for aquatic and wetland plants, whose adaptations aren't necessarily to local areas and climates, but to a wider adaptability when carried by waterbirds or downstream flows to new places. This flexibility of response is why many aquatics are among the most naturally widespread plants in the world, and humans have now made it possible for diverse aquatic weeds to cross oceans which were once an insurmountable barrier to their spread.

The problem is made worse because Australia is not one climate but sprawls from cool temperate zones to the tropics, and plants that show little weed potential in any one area may become serious problems if introduced elsewhere. On the positive side, it should also be theoretically possible to control potential new weeds across the whole of this considerable landmass, as long as all states and territories are prepared to impose blanket bans on any plants that may become weeds if introduced into a more suitable climate. There are many such potential weeds still being legally sold in Australia, and restricting or prohibiting them would be made easier if gardeners could be shown that few of these are in much demand, and that their economic value is negligible, yet the escalating price for control if they escape will ultimately come out of the pockets of taxpayers and ratepayers.

Problems caused by weeds

The problems caused by weeds are not necessarily the fault of the plants themselves, which are often growing out of control because of environmental factors, particularly in the many disturbed habitats humans have not only created but continue to maintain for various specialised purposes. Although some agricultural weeds (particularly of drains and rice fields) are discussed in this book, these are primarily colonising species that thrive best in disturbed conditions. It is environmental weeds that are of most concern because these can invade relatively undisturbed wetlands, displacing indigenous plants and altering natural habitats and processes with further loss of habitat for native animals.

Other weeds significantly increase evapo-transpiration, as they take up and release water through their leaf surfaces (transpiration) at a greater rate than evaporation alone can achieve. Willows and poplars (see plate 25) are particularly thirsty, few indigenous plants will grow under them, and as they shed all of their leaves over a short period in autumn they deplete dissolved oxygen levels and turn streams into impoverished ecosystems that support only a limited range of animals. Their brittle timber snaps easily in storms, and their fibrous root systems are easily torn out of the ground during floods, accelerating the pace of erosion.

Other wetland weeds provide habitat for pest species from insects and snails to feral pigs. They can interrupt or slow natural flows so that sediment builds up faster, and some species that thrive in deeper cooler waters may clog or interfere with the working of weirs and hydroelectric plants. They may also prevent boating, fishing and other recreational uses, or taint water so that even livestock drink from it reluctantly, and reduce yields of fisheries. All of this comes with the ever-increasing monetary cost for control and management, because if they are left to spread, or new weeds are left unchallenged, both wetlands and waterways will become increasingly useless as habitat or for food production.

Causes of weediness

Rather than looking at the way a plant grows for an immediate and possibly simplistic explanation of weedy behaviour, in many cases we should be looking first at the base conditions, often created by humans, under which it has become weedy. High nutrient levels from agricultural, residential and industrial runoff are the most obvious example of this, and many plants which grow profusely when over-fertilised may be quite mild-mannered in cleaner waters just a few metres away. Rice fields (see plate 30) create comparable conditions, because rice is an annual weed that happens to produce a useful crop. These flooded, fertilised fields with few competing plants also attract a wide range of other annual weeds, many of which set their own crop of weed seeds much sooner than the rice itself.

Nutrients are concentrated in wetlands by definition, because these low-lying places are where water naturally accumulates or passes through. Unnaturally high concentrations of nutrient also create other problems which the plants may be blamed for, such as algal blooms which reduce dissolved oxygen levels, stench from anaerobic conditions (sulfurous-smelling mud is particularly common),

An ornamental pond fed by nutrient-rich, urban runoff, creating dense growth of diverse weedy aquatics including parrotfeather, cumbungi, Pacific azolla and willows; as a result silt and plant material have filled the pond with detritus so that it is now mostly less than 1 metre deep.

disappearance of most aquatic animals from the vicinity, and sometimes even dead waterbirds during times of low water when the birds are feeding in exposed mud containing toxic bacteria.

Although hardly pristine environments, urban waters are much cleaner these days than they were a few decades ago. Most of the pollution, toxic waste and dissolved nutrient affecting wetlands and waterways probably now comes from agriculture, and even trace amounts running off vast acreages can have significant effects when concentrated in a sluggish stream. Although the Land Care movement has made significant reductions to nutrient runoff, there are still thousands of kilometres of waterways that are crying out to be replanted with indigenous species, set into cut-off drains or swales to allow nutrient absorption by terrestrial plants before water enters nearby wetlands.

Regardless of how clean and low-nutrient water may be, there will always be plants that grow in it, and in many situations these now include weeds. As a group, aquatic weeds are characterised by rapid vegetative growth, many of them regenerating from small or even minute fragments, or dormant organs such as tubers and turions. Colonising species are even more adaptable and invasive than aquatic weeds, as they are adapted to travel between wetlands either as wind-blown seed, or attached in some way to migrating waterbirds. Arriving on disturbed or uncontested ground, such as recently exposed lake floors, such seeds may germinate in countless billions.

The unpredictability of wetland weeds

Despite many decades of study, categorisation schemes of sometimes confusing degrees of complexity and diverse (and sometimes contradictory) concepts supposedly helpful in predicting potential weediness among aquatic and wetland plants, the best guide to weed potential we have to date remains practical experience. This is technically known as 'learning the hard way', because escaped weeds are usually difficult to put back in their box. Even on the basis of practical experience it is difficult to extrapolate from one situation to another, as to how any particular plant is likely to behave in a new set of circumstances and growing conditions.

For example, mud plantain (*Heteranthera reniformis*) is regarded as endangered in parts of its native North American range, is seen as a serious rice field weed in Italy, and is also still legally sold as an ornamental – though introduced – plant in Australia. Here we have an example of humans potentially creating a serious weed through a combination of artificially created and maintained habitat, juggled unskilfully against the perceived desires of a large and influential market made up of keen gardeners, whose wishes must not be thwarted.

However, from discussions with some of the longer-established water garden nurseries, it seems likely that only a few dozen plants of this potential weed are sold annually *through the whole of Australia*, and the princely return of several hundred dollars a year is shared among those few nurseries which bother to sell this uninspiring weed. If surplus plants of this species should be dumped into an irrigation channel, and it spreads to rice fields, the ultimate cost of its control and eradication in Australia is likely to be in the millions of dollars – assuming that it can be contained at all.

Biological categories of wetland plants

Systems of categorising aquatic and wetland plants can help with understanding their basic biology and growth patterns. However, they should ideally be kept as uncomplicated as possible because both the plants themselves, and the diversity of unnatural habitats made possible by humans, create many exceptions. The more complex an ecological category, the more special cases have to be made. For the purposes of this book all wetland weeds are grouped together as *free-floating*, *submerged*, *rooted with floating leaves*, *emergent*, and *water's edge* species. Algae make up a very different set of categories, but as most of these are difficult for the non-specialist to identify, and these diverse organisms already have an extensive and specialised literature of their own they are only considered briefly at the end of this book.

Plant zones

Wherever wetland plants grow, they must have adequate supplies of light, carbon dioxide and oxygen to carry out their essential business of photosynthesis, which is basically fixing solar energy into energy-rich compounds that also happen to be the ultimate source of food for nearly all animals on earth. Each type of aquatic plant has its own preferred zone where it is most efficient, or better able to tolerate the difficulties and seasonal changes in conditions than other species. What humans may sometimes mistake for weedy growth is sometimes just an indication that a particular plant is better suited to a difficult environment than any other species present, so that it can grow without significant constraint or competition.

Water is low in dissolved gases and is also often murky, which is why so many wetland plants strive to reach the water's surface where light, carbon dioxide and oxygen are more freely available. Some are able to change their growth habit according to circumstances, for example, floating plants that can also take root in exposed mud, or emergent species such as *Sagittaria* and *Sparganium* (see plates 30 and 7) which grow as more streamlined, submerged forms during periods of flood, or in permanently deeper waters.

Dissolved carbon dioxide is a significant limiting factor, and it is likely that many fully submerged plants in clear water would grow many times faster if more of this gas was more readily available to them. Availability of oxygen to the roots is a further limiting factor, though many fully submerged plants only have small root systems and can obtain most of their needs directly through the surface of their leaves. Submerged and waterlogged soils are particularly low in oxygen, or may even be anaerobic, so the hollow stems of many emergent water plants are designed to move oxygen down to the roots.

Free-floating plants

Floating plants mostly drift on the water's surface, moved freely by wind and currents so they are most abundant in relatively still or slow-moving waters. In most such species the roots trail below the water's surface, though in the smallest duckweeds there may be no roots at all because the plant can absorb all it needs from the water directly through its surface. Of the indigenous floating plants, most duckweeds and the two species of *Azolla* (see plates 15 and 16) are sometimes regarded as weedy in high-nutrient situations, while the introduced water hyacinth (*Eichhornia*) and *Salvinia* (see plate 17) were regarded as being among the world's most problematic weeds until relatively recently.

In still waters with high nutrient levels, some floating plants can double in area within a week or two, ultimately creating a smothering blanket that reduces oxygen supply to the water below, and also reduces light levels so submerged plants and the phytoplankton which forms the food base of many aquatic ecosystems disappear. In turn, that means most carnivorous animals that would normally control populations of mosquitoes and other disease-bearing pests may become rare or disappear altogether, so a dense carpet of any floating plant can have a dramatic impact on underwater ecology. As some of these plants also die off in large numbers under adverse conditions, the decay they create below depletes dissolved oxygen still further, and taints it so it may be unfit for human or livestock consumption.

Submerged plants

Plants that grow fully submerged at all times may send up tiny flowerheads above the water's surface, but otherwise they have to be a tough breed able to survive months of murky water, and even floods, by slowing down their pace of life. Most submerged plants attach themselves by roots in the mud below, one widespread exception being hornwort (*Ceratophyllum demersum*; see plate 13), which may tangle itself around other plants or grow in deep, clear waters attached by specialised structures that are not roots but anchors. Other submerged species almost belong in the next ecological category to be described, for example lake watermilfoil (*Myriophyllum salsugineum*) which grows mostly submerged, but waves a tiny, emergent flowering stem with small leaves above the water in the warmer months.

Rooted plants with floating or slightly emergent leaves

Many aquatic plants grow in moderately deep waters, yet concentrate most of their effort into producing leaves that reach the surface, or can grow above it. Such plants are often split into many other categories in more complex systems, for example in one system 'euhydrophytes' which may have submerged and floating leaves at different stages. None of these further categories seems to have found wide acceptance, so we can leave their originators to split hairs for the moment, and look at how the plants actually grow. Some such as floating-leaved pondweeds (*Potamogeton* species) have a specialised underwater leaf that dies away as soon as it reaches the surface, to be replaced by rounded floating leaves that receive full sunlight and a more reliable supply of both oxygen and carbon dioxide. Weedy floating-leaf plants such as starwort (*Callitriche*; see plate 27) may grow so densely as to form a floating mat, and in extensive shallows can reduce dissolved oxygen levels almost as effectively as some floating plants.

Emergent plants

Many aquatics grow rooted in mud, with most of their stem and leaf tissue held above the water's surface, giving them free access to both light and the gases they need to grow. Often referred to as macrophytes, these include the most varied wetland weeds, from sedges to water primroses and arum lilies. In high nutrient situations where everything they need is freely available to them, these are also among the most vigorous and invasive plants, and many of them are capable of penetrating even into relatively undisturbed natural wetlands.

Water's edge plants

Above normal high-water level soils are only flooded intermittently, so for most of the time the air spaces within the soil allow less specialised plants to grow, although even these must be able to survive the reduced-oxygen conditions during periods of flood. Water's edge plants cover a wide range of species from sprawling creepers to tall trees, and those growing along stream banks where floods are common but unpredictable are referred to as riparian plants. Not all species growing near streams are wetland plants, and some significant streamside weeds including lantana (*Lantana*), blackberries (*Rubus*) and willows (*Salix*) are briefly discussed later (see plates 24 and 25), as these are often the dominant species along many kilometres of streamside habitat.

Legal and official categories

There are numerous schemes used to classify and assess aquatic weeds in Australia, ranging from largely agricultural classifications such as 'noxious' – which may be applied even to useful habitat species, through various categories and threat levels, often built around elaborate numerical systems that are not necessarily easy to interpret. These vary from state to state as well as at the federal level. Even within a single state the various authorities and government bodies that may overlap in responsibilities don't necessarily coordinate their efforts and classifications to produce a single, easily understood system.

Although there is a considerable amount of information available on at least some of these weeds from diverse government authorities and on the internet, for anyone coming in cold from the outside, seeking to identify or understand the potential of aquatic weeds in their area, the overall impression must be confusing. The diverse categories used include such terms as State Prohibited Weeds and State Alert Weeds, while others are escalated to National Alert Weeds, National Sleeper Weeds, and some (but *only* some) of the worst of them are Weeds of National Significance. Rather than reproduce the explanations of these and many other categories, a guide to the websites where they are defined and explained is included along with other references used for this book.

Classification into one or another of these categories does not necessarily give a true picture of just how weedy any one species may be. For example, the Weeds of National Significance (WONS) list includes precisely 20 species that have been chosen as the defining weeds of our time, though this category is revised every few years. While the overall principle is good, a simplification of this kind may inadvertently give the impression that anything not on that 'most wanted' list is comparatively benign.

It is not surprising that the current WONS list includes quite a few wetland plants such as alligator weed (*Alternanthera*), cabomba (*Cabomba*), olive hymenachne (*Hymenachne*), pond apple (*Annona*), salvinia (*Salvinia*) and most of the willows (*Salix*), other than a few species which the various states couldn't come to any agreement upon. Yet even if this list was to be increased to 50 species, there would be no difficulty in making up the numbers for a list of seriously threatening plants, including many other wetland species.

The problems and confusions spread by this multi-layered, sometimes regionally biased and far from consistent approach to dealing with wetland weeds are compounded by the proliferation of information on the internet, not all of it accurate; see *Ludwigia*, for example. And somewhere under the radar fly many species that are still readily available as ornamentals because no one seems willing to tackle gardeners and the lucrative garden market, though some of these have real potential to become a new generation of wetland weeds in the not-too-distant future.

There is no possible way to join these various and diverse stages and categories of weediness into a single tidy scheme, so that anyone (and particularly gardeners, who as a group are the single most likely vector for the spread of new weeds) trying to work out just how threatening any particular plant could become, will most likely be left confused and unenlightened. For example, the ageing term 'noxious' covers a multitude of sins. It is generally applied only to plants that have already well and truly bolted, including a number of indigenous species simply because they may become a nuisance under some conditions.

Thus the two native species of *Typha* are considered noxious in places along with the introduced cat-tail (*T. latifolia*), and could therefore be interpreted as equal in weed value (see plates 2 and 3). Yet any attempt to control, let alone eliminate either of the two abundant and widespread native species would use up a significant proportion of the gross national product, and would also have to ignore their real (if limited) value as habitat. By contrast, cat-tail continues to spread by wind, appearing in unexpected places, yet is largely ignored by local councils and wetland managers because the present categories of weediness don't differentiate it from its indigenous relatives.

Nor does a plant have to be introduced, toxic, or even invasive to be spuriously included in the most comprehensive lists of 'weeds'. One extreme, though admittedly non-aquatic, example is the native yam-daisy, cited as *Microseris lanceolata*. This once abundant and ecologically significant plant of native grasslands (now regarded as a group of closely related and comparably endangered species) was an important and highly regarded food for many Aboriginal peoples, but has largely disappeared as it is preferentially eaten by sheep and other livestock. Indeed, the only justification for including it in a recent catalogue of weeds appears to be that it has managed to persist in some overgrazed pasturelands, despite efforts to make the pastures less diverse in species!

At this most extreme edge of the spectrum, a recent learned tome of introduced species being grown in Australia includes tens of thousands of plants, with the barest minimum of information as to whether any of these have ever actually been reported as weedy anywhere, and no clear indication as to the criteria used for inclusion apart from numerous unreferenced letters and numbers. Despite its avowed claim to help its audience (it is aimed at diverse groups from researchers, farmers and consultants to gardeners) 'make informed decisions about which [plants] to use in their operations', this all-inclusive approach is so uninformative that it could lead to the blacklisting of virtually every exotic plant in the country, and is likely to alienate most of the groups whose support it is attempting to enlist.

This proliferation of definitions and classifications is used by a wide range of wetland managers in different ways, creating still further confusion. Thus native plants in rice-growing areas may be regarded as weeds while their siblings on the other side of a canal, in a natural wetland become habitat plants. And even within one wetland managed under very different regimes or authorities, cumbungis (*Typha* species) may be seen as useful habitat on one side of an arbitrarily painted line, and an obstruction to water flow and quality on the other.

Rather than cobble together yet another classification system for aquatic weeds, each individual plant in this book is discussed in the context of how it grows and spreads, its impacts (real or imagined) in undisturbed wetlands, and whether there is any prospect of containment or if it is just too late to even try. This should allow readers to assess the potential impacts, dangers, and likelihood of successful control in their own area and situation. In some cases, such as introduced waterlilies (*Nymphaea*), which involve a wider range of issues including public perceptions of hidden problems, the presence of sometimes cryptic indigenous species, and the specific attributes of particular introduced species and cultivars in differing climates, the discussion is wider-ranging and has been formatted accordingly.

Weedscapes

Weed-dominated landscapes (referred to as weedscapes in this book; see plate 1) are often attractive to people who know little about them, creating an impression that all is well and in balance even in the most disturbed environments. One of the greatest problems facing attempts to control aquatic weeds is that most people, particularly in urban areas, don't realise that some of the species they may regard as a desirable component of the landscape are actually weeds that adversely affect habitat values, water quality, water availability, and even their taxes: the latter piece of information would probably hurt more than anything else!

For two centuries we have been importing new plants as grazing for livestock, for visual appeal, to recreate the appearance of places our forebears were nostalgic for, and to bind the banks of rivers that were held together perfectly well by the native vegetation we cleared. Thinking of the superficially bucolic, English feel that willows lend to the landscape, I am all too aware that the streams that run beneath them are biologically sterile compared to those same streams 200 years ago. And over the past decade of drought in south-eastern Australia I have also become increasingly aware that these water-hungry plants are significantly reducing the water available not only for environmental flows, but also for agriculture, which will become increasingly critical for human survival if even the mildest climate change scenarios are proven correct.

Education must be the key element in changing people's perceptions of the contemporary landscape, and in particular their perception of weedscapes, but in an era of information glut it is hard to see where to begin. The numerous and diverse problems caused by weeds just aren't sexy, even though a new generation of potential wetland weeds (along with their associated environmental and economic costs) is still being legally sold through nurseries. Weeds aren't our friends, nor are they out to get us, but there is an urgent need to come to grips with wetland weeds while there is still some prospect of success.

2

Prevention, control and management

The primary theoretical defence against weeds of all kinds is prevention through education, yet few people have any real knowledge of the environmental and economic costs of weeds, and nor are they in any hurry to learn about them. Ironically, the abundant information on the internet not only gives an impression that weeds (both terrestrial and aquatic) are being adequately dealt with, but is also so dispersed, voluminous and sometimes contradictory that it is also off-putting to anyone with just a casual interest. Added to this there are numerous wetland weeds that are so localised in their impacts that they don't produce much of a blip on the radar of public perception.

The overload of information created by a superficial internet trawl can also create an impression that there are many more serious wetland weeds out there than we have any real prospect of controlling. This misleading impression is compounded by the descriptions of some relatively minor wetland weeds, which may be described in the same grim terms as the major, fast-spreading species. Yet many wetland weeds have been naturalised for a very long time in Australia and remain relatively localised, or may not be spreading at all. There are even some cases of weeds, which have probably already occupied most of the possible range available to them, where there is little evidence that they have caused significant ecological harm.

It is important to be able to differentiate between the various types and levels of threat posed by the introduced weeds and potential weeds described in this book, because fighting against many of these would be an exercise in futility, at least with the knowledge, technologies and funding levels available at the present time. It is far more useful to concentrate on the control of weeds that could spread more rapidly and widely yet are still potentially containable at the present time, and to be able to differentiate between seriously invasive species capable of invading relatively undisturbed wetlands, and plants that are only notably weedy in disturbed places such as open drains, irrigation canals and rice fields. A further group that needs consideration is those weeds that at present are only declared *non gratis* in some places, yet will undoubtedly become increasingly abundant and widespread if they are not tackled at the national level.

These and other related issues are looked at in the discussions of individual weeds which make up the bulk of this book. No attempt has been made to create yet another artificial framework or numerical ranking of their possible impacts. Instead, each species is discussed in such a way that readers can make up their own minds as to how potentially weedy any one species may be (or may become) within their local wetlands. In terms of the practical management of local wetlands, say within a 50-kilometre radius or so, we usually only need to consider around one or two dozen established weeds out of the 130 or so introduced species covered in this book.

In turn, a realistic appraisal of the threats posed by locally entrenched wetland weeds also makes it more feasible to educate other people in your area about their impacts and potential impacts. The emphasis is on making sure new species don't have a chance to establish from garden waste, aquarium dumpings, or being brought in with heavy machinery. This should be the primary purpose of education, yet it can fail in unexpected ways. Consider the recent rapid spread of alligator weed (*Alternanthera philoxeroides*; see plate 28); who could have predicted the spread of this species as a popular food plant by Asian communities? Appropriate educational programs have hopefully stopped its further spread, but in the several years before these took effect, alligator weed had changed from a regional problem in the central New South Wales coastal region, to a fast-spreading weed in many urban areas from Brisbane to Melbourne.

Education is an important potential tool in controlling the spread or introduction of introduced plants into natural waters, but will never be enough in itself. Perhaps there will always be a few people who continue to regard dumping of surplus pond (or garden) plants as being in some way a humane means of disposal, or who are stupid, or just don't care. It is likely that we can contain and hopefully correct the damage such people do from time to time, if we aim education campaigns at the two demographic groups which are most likely to introduce new weeds: gardeners and aquarists.

Aquarium and pond plants

Some of the most serious wetland weeds have been introduced as pasture grasses, as erosion controls or for 'rehabilitation', and occasionally for their edible properties, but most notably they have been introduced by the garden and aquarium industries.

The aquarium and pond plant trades are just a minor part of the garden trade, itself a significant source of new weeds both aquatic and terrestrial, yet these two specialised trades have been responsible for the introduction of most of the worst aquatic weeds in Australia. The actual gross, economic value of these specialised trades in Australia, and the very restricted market they support, is surprisingly small when weighted against the considerable costs for control of aquatic and wetland weeds which have been distributed through these markets: probably not much more than two million dollars per annum.

Of these two specialist areas, the pond or **water garden** market is the larger, although it still only makes up a tiny fraction of the nursery trade within Australia. This is because relatively few gardens have ponds or other water features, and such features usually make up only a small portion of the garden. Seen in this light, as a minority hobby catering to the interests of perhaps one in a hundred gardeners, the number of serious weeds that have been introduced by the water garden trade is impressive. However, it should be kept in mind that the wider nursery trade is also still introducing new and potentially fast-spreading wetland weeds, and legally importing a number of inadequately assessed species in the form of seed.

The water garden trade continues to offer a number of potential weeds, which have been made freely available over a wide range of Australia's climates to date, although many species are already prohibited in at least some states with very good reason. This means gardeners knowing nothing about the weed potential of many of these plants could be inadvertently testing them in diverse Australian climates and conditions, including in places where they may have a better chance of flowering and successfully setting seed than at present. Most of these plants have yet to spread into natural waters, but continued introduction of new strains or clones may make it possible for even some of the most infertile clones being traded to ultimately spread between wetlands by seed.

The 1990s fashion for water garden plants has largely passed, and competition between water garden nurseries to introduce novel varieties or clones of aquatic plants from overseas has decreased over the past two decades. In part this is because of increasingly stringent biosecurity measures, including the ever-increasing cost of bringing in allowed novelties through quarantine. The limited market appeal of (expensive) recent introductions means that in most cases the import costs cannot be justified, except for higher value plants such as waterlilies (*Nymphaea*; see plate 18) and lotuses (*Nelumbo*).

Many of the older waterlily varieties and cultivars in Australia are known to be relatively infertile and non-invasive, but the varieties which generate the most excitement overseas (and therefore seem most desirable to local importers) are often more vigorous selections of species, and recent primary hybrids. These are popular with overseas breeders because they are generally much easier to breed from, unlike many older hybrids and selections which are largely or completely infertile. In other words, not only are the more recent *Nymphaea* cultivars most likely to be imported, but they are also the forms most likely to set fertile seed. This is a reality that should perhaps be considered in the future assessment of waterlily cultivars suitable for importation.

Smuggling of prohibited species has probably diminished considerably since the 1990s, but some new water garden and aquarium plants continue to appear from time to time with little indication of where they arrived from. During the course of my experimentation with exotic aquatic and wetland plants, I catalogued every species being grown within Australia, using information freely provided by diverse water garden nurseries and by private collectors. Although I stopped updating this list in 2002, Hibbert's 2004 compilation of garden plants and sources was a useful confirmation that no new aquatic or water's edge plants had been offered for many years.

Despite increased inspection and a greater emphasis on biosecurity in the decade since the terrorist attack on the Twin Towers, new introduced plants have appeared on catalogues since that time, for example the dubiously identified *Alisma parviflora*, *Carex grayi*, *Cyperus percamenthus*, new horticultural forms of *Juncus effusus*, *Limnobium laevigatum*, *Lycopus europaeus*, three *Micranthemum* species (two as *Hemianthus*), *Samolus parviflorus*, and two distinct forms of *Shinnersia rivularis*, a daisy with comparable weed potential to its relative *Gymnocoronis* – except that it is even more cold tolerant!

Yet in practical terms, water garden nurseries don't need to keep importing and offering hundreds of exotic aquatic and wetland species, weedy or otherwise. Most garden ponds are only a small part of the garden, and their owners are unlikely to want more than a dozen or so species or cultivars for an initial planting, though they may add another plant or two in later years. Most such plantings would usually include several waterlilies, and perhaps some irises, of which dozens, if not hundreds of relatively non-weedy cultivars are available. To supplement this basic range most gardeners are content to choose from a basic range of proven non-weedy exotics that are reasonably showy in flower, and even general nurseries stock an adequate range of these.

It is far more likely to be collectors (including some nurseries) who obsessively accumulate all the exotic water plants they can lay their hands on, regardless of whether these are particularly ornamental, or their potential environmental impacts. Here we are talking about perhaps one or two hundred people in the whole of Australia. There can be no economic justification for continuing to allow the sale of an extensive range of potential wetland weeds, when a great many of these species return *gross* sales well under a thousand dollars per species, per annum, through the whole of Australia.

By contrast, the cost of even the most basic control program – should some fool dump unwanted plant material of such novelties near a river or wetland – is likely to be in the order of hundreds of thousands of dollars. If all of the potential water garden and aquarium weeds described in this book were banned outright from sale or trade, I estimate the cumulative financial loss (spread over *all* garden centres, water garden and aquarium suppliers in the country) to be well under $50 000 per annum. Yet if even *one* of these species naturalises, the control measures for that single species are likely to be much greater than this, even assuming that complete eradication can be achieved within a year or two.

For this reason the range of introduced aquatic plants should be reduced to a clearly defined list as has long been the case for fishes imported from overseas, with new additions to the list undergoing a detailed review process before being accepted – in other words, deemed to not be a significant potential environmental or agricultural weed. The only reason this has not been done for water plants is simply a matter of politics; gardening is big business, operating on a much grander scale than the aquarium fish trade, so politicians and government departments are reluctant to come to grips with it.

The present situation could be compared to the years where tobacco smoking was condoned as it brought in an apparently generous tax income for government, but was gradually curbed and restricted. The reasons for this were partly economic, as it was eventually realised that *all* taxpayers were funding the ever-increasing medical bills resulting from tobacco use, and these were much greater than the tobacco revenue could justify. In the case of potential wetland weeds, the escalating cost of weed control is already a good enough reason to stop all trade in potential weeds, as the long-term costs of those that have already gone feral will be a perennial burden, and we don't want to add to it.

Aquarium plants are a still more restricted range of those aquatic or semi-aquatic plants that can survive or even grow in extremely low light, although they rarely flower under these conditions. This is the most specialised form of water gardening, requiring a considerable understanding of lighting and fertilisation, as well as the role of dissolved gases including carbon dioxide for best growth in the artificial confines of the aquarium. There are probably only a few hundred serious aquarists in the whole of Australia who can grow such plants well, and as these people almost invariably have a keen interest in native plants and wetlands as well, there is little danger that they will dump surplus plant material into natural waters.

The main danger posed by aquarium plants is those adaptable species that survive even inexpert attention, so that if they grow too well surplus materials may be dumped into rivers and wetlands rather than be destroyed, as an act of 'kindness'. Many of these species spread readily from broken pieces, rhizomes and other smaller fragments even if they don't flower or set seed, and some have even been deliberately planted by people who were planning to harvest aquarium plants from the naturalised stands, until they turned out to be so weedy that they were prohibited!

It would not be difficult to produce a simple and attractive poster warning of the environmental dangers of dumping, and of the counterproductive nature of deliberate introduction. Some posters relating to wetland weeds have already been made, though they emphasise only the major species, and it is not easy to persuade aquarium shops to display them prominently. A greater awareness of indigenous alternatives for many of the exotic plants offered by nurseries would reduce demand for potentially weedy species further, though the issue of how widely such 'native' species should be available is discussed in the next chapter.

For a more detailed discussion of weeds and potential weeds introduced or distributed via the aquarium and water garden trades see: *Aponogeton*, *Aristea*, *Bacopa*, *Cabomba*, *Echinodorus*, *Egeria*, *Eichhornia*, *Elodea*, *Epilobium*, *Equisetum*, *Gunnera*, *Gymnocoronis*, *Heteranthera*, *Hydrocleys*, *Hydrocotyle*, *Hygrophila*, *Iris*, *Isolepis*, *Lagarosiphon*, *Ludwigia*, *Mentha*, *Myriophyllum*, *Nuphar*, *Nymphaea*, *Pistia*, *Pontederia*, *Rotala*, *Sagittaria*, *Salvinia*, *Shinnersia*, *Typha*, *Utricularia*, *Wachendorfia* and *Zantedeschia*.

Assessment and planning

Wetland weed control needs to be planned within the overall management goals and aspirations for any type of wetland, in the context of the type of habitat certain weeds provide or destroy, their effects on water quality, and whether they can be replaced with more appropriate indigenous plants successfully. Unfortunately, in many wetland situations it is not clear which organisation or government body is in charge, so there is often no overall, coordinated plan for management. This is partly because many official bodies sensibly wish to avoid taking direct responsibility for a wetland or section of river unless it is already clearly within their brief.

Accepting new responsibilities of this kind doesn't mean that additional staff or funding will be automatically made available to deal with the additional workload, but more usually results in additional strains on resources which are often already inadequate. And in urban situations especially, anyone who cares enough to shoulder the responsibility for a degraded wetland or waterway may also be subjected to criticism from diverse special interest groups, regardless of whether such groups are prepared to actually carry out any of the necessary work. To deal with such problems in our increasingly bureaucratic times, advisory committees are likely to be set up, and anyone who has had experience of these will know that hundreds of hours can be wasted in fighting and bickering over divergent agendas, and sometimes even the opinions of particularly vocal individuals.

As a result, cleaning up wetland weeds sometimes becomes a type of free-for-all, where any group willing (in good faith) to tackle a problem may find itself in a position where it may be granted funds, without being given adequate time to assess the situation. I was recently approached by a group that had been granted funds to spray a weed which has too long been neglected in a nearby river, with a simple question – should they spray now, or in spring? The grant had been made without assessment of the impacts a sudden die-off of large areas of the weed would have on the health of the river, whether there is a significant buried seed bank that would germinate rapidly to take advantage of the newly disturbed habitat, what sprays were appropriate, or even if the weed in question could realistically be controlled by spraying with anything legally available.

In the case of a recently arrived weed which is still to establish itself on any scale, urgent action is often appropriate, and may be as simple as manual removal with follow-up removal of seedlings or fragments which remain. Such an action is unlikely to be in any way controversial, but in the case of any well-entrenched weed problem *assessment is the essential first step*, however long this may take, followed perhaps by small-scale trials of proven or likely control measures before a longer-term plan is considered or reaches the stage of public input.

Any such program should address (at the minimum) key issues such as:

- the actual impacts and likelihood of further spread of the weed
- whether there is a long-lived seed bank ready to germinate when the parent plants are killed off
- whether there are other and possibly still less desirable weeds ready to take its place
- the effects of any spray program on the wider environment including on habitats that are purportedly being adversely affected by the weed; and
- what types of indigenous plantings should be used to try to achieve long-term control, overshading or replacement of the weed.

Yet, as I emphasise throughout this book, it is not always possible to control weeds, often because there will never be enough funds to finish the job once and for all, but still more often because it is just too late to start, or no viable control method for a given situation is known. Some weeds may not need to be controlled, but only managed so their impacts are reduced to an acceptable level, while others are already so widespread and well-established that they are likely to reappear in places where they have been locally eradicated. If a comprehensive assessment has been made and it is clear that there is a real prospect of eliminating, or at least containing the spread of a weed within a defined area, with positive environmental outcomes, the physical work can begin.

Restricting further weed spread

Preventing further spread is the first tier of physical defence against wetland weeds, and elementary care such as cleaning seed or viable fragments of aquatic weeds off machinery, boats and other equipment should be routinely practised when carrying out any kind of work in a wetland where a weed is already established. This *particularly* applies to machinery specifically used for this task, although some operators have been known to take a fairly cavalier attitude to this in the past. Standardised procedures for such precautionary cleansing are widely available on the internet, and are also routinely included in many of the Control Manuals cited on the Weeds of National Significance website (see Recommended Reading), available both online and in hard copy.

Routine sterilisation of *any* equipment used in one wetland before it is used in another will also help to reduce the spread of some diseases, not just of plants but also the ever-increasing range of imported diseases that may affect indigenous animals. Even protective clothing such as waders and wetsuits, nets and collecting containers should be treated: I hose any mud off mine in a location where no runoff can reach permanent water bodies, allowing everything to dry out completely before reuse, and even spray with a dilute disinfectant solution before drying if I have been working in a suspect location.

A further precaution against spread of wetland weeds is minimisation of disturbance, especially where there is a dense and established cover of indigenous plants. Many weeds are colonisers of bare spaces, so clearing of any kind should be avoided as it creates openings which the fastest-growing plants are likely to occupy ahead of most other plants – that's why we regard them as weeds! Even if such entrenchments are small and can't spread further because they are surrounded by indigenous species that can hold their own against the invaders, established patches of weeds are likely to produce seed or other material that will be quick to colonise any further disrupted or damaged areas.

Control methods and combinations

The various control methods described below are often more effective when used in combination, especially in a well-timed sequence, ideally taking advantage of natural seasonal changes in growth to weaken weeds as rapidly as possible. As one example, deep cutting of submerged aquatics will be more effective if followed up by increasing shading or turbidity. The remaining rootstocks in deeper waters (where light penetration is already borderline for reasonable growth) will already be under stress, so any further decrease in light levels could become life threatening if it can be kept up for at least a few weeks. The minimisation of seed production should also be kept in mind as a priority when planning a series of actions against any particular species, except for the handful of introduced weeds which can't set seed.

Even the use of chemical treatments is often more effective when applied after cutting, altering soil or water pH, grazing, or any other disturbance that has already made inroads on the growth of the weed. The removal of a significant volume of any weed by physical means before applying any toxin not only reduces the actual quantity of toxin needed, but also makes it possible to more directly target the remaining growth, including the crowns of grassy weeds.

Manual and other physical control methods

Manual control methods should always be considered as the first field of engagement with any wetland weed, particularly if it has only just started to show its presence. Any measures taken to remove a small number of individual plants at an early stage have a real chance of preventing the building-up of a soil seed bank, which would later become a perennial source of re-infection. In the early days of an infestation it is often possible to remove every plant of a newly arrived weed by hand. Even where a recently arrived weed has been detected in a number of places in a wetland, staggered removal concentrating on destruction of the weed in more vulnerable or disturbed areas may be a viable option, reducing seed formation in the places it is most likely to germinate. The same weed growing in more crowded and competitive situations can be left to be tackled at a later date.

For larger infestations, slashing with a tractor will weaken even established weeds and reduce the chances of them setting seed. In wetland situations, this is only practical in ephemeral wetlands or fringes where the soil dries out enough to allow machine access at times. Some weeds have a very limited window of opportunity for flowering, so cutting them when they already have much of their energy invested in the flowerheads which are forming not only prevents any seed being produced in that season, but also sets back their growth more dramatically than at other times when more of their reserves are stored away in less accessible parts of the plant – such as in the rhizomes.

As a short-term measure applied over several consecutive years, slashing (or any comparable damaging process including the use of a brushcutter on smaller infestations) may make it possible to eventually tackle the complete local eradication of any particular weed without having to consider the possibility of germination and regrowth from a buried soil seed bank. For example, a plant that has a likely maximum seed viability period of three years will have its seed bank severely depleted by cutting for three years running, though some allowance may need to be made for seeds that may survive longer if buried in mud, than in drier situations.

For annual plants, which basically flower, set an abundant supply of seed, and then die, slashing or using a brushcutter to remove all flowerheads before seed is formed may resolve the weed problem in a single season. However, annual weeds are also generally more persistent in their attempts to set seed as their survival depends on it, so even the most enfeebled annuals may flower repeatedly in an attempt to set at least a few seeds before their time is up. Cutting timed to prevent tuber or turion formation in species that depend on these dormant storage organs to survive during dry or cold periods will also reduce the chances of them reappearing in the following season.

For some non-toxic weeds on agricultural land, grazing (or slashing followed up by grazing) may be an even better option for control, as evidenced by the common reed (*Phragmites*; see plate 7). Although only briefly considered as a weed in this book because it is both indigenous and a significant habitat type in its own right, this tall and hard-stemmed grass often forms dense, almost impenetrable thickets on muddy ground and in shallow water that may cover many hectares. Such thickets give a strong impression that they would be almost impossible to manage, yet cattle eat the young shoots with relish, and in many places dense stands of the reed disappear completely on the other side of a fence where cattle are present.

Less well understood is the effect of sudden changes in water pH, which has been used with some success to eradicate aquatic vermin such as unwanted fishes in *small* bodies of water. Many aquatic plants prefer relatively acid water, and some have only a limited tolerance for lime or other alkaline substances. Increasing water pH with a bombardment of lime or dolomite at a strategic time – such as just before flowering – can have significant effects on the production of seed, though it will also adversely affect many aquatic animals as well. This method works particularly well on floating plants that have no contact with the submerged soils below, but is less reliable for rooted plants as many of these are more strongly affected by soil pH than by the condition of the water.

Harvesting of some weeds, and not just edible species, has sometimes been suggested as a commercially feasible control measure. Even water hyacinth (*Eichhornia*) has been dried and woven into mats and furniture, composted, and also shredded and used as a mulch in parts of Asia. Although the work involved in this kind of weed control is never-ending, it is also a source of considerable volumes of organic matter free of terrestrial weed seed. Also, with low intrinsic nitrogen content, it tends to break down relatively gradually when used as a mulch. The harvest of floating weeds (such as *Eichhornia* and *Pistia*; see plates 17 and 16) could be a relatively passive process in rivers and other moving waters, using some form of diverting barrier to redirect floating mats into a pocket near the shore, where they could be mechanically harvested.

However, use of this kind assumes that harvesting is only done in fairly clean and pristine waters, as many weedy aquatics accumulate heavy metals and other pollutants. If established as an industry, this form of weed management can potentially become the cause of legal actions and disputed priorities in weed control, as any industry that depends on a single weed as its primary source of income will automatically be against the further introduction of biocontrols, though in reality there is not a single wetland weed we wouldn't be better off without.

In wetlands where it is possible to control water levels, drawdown or flooding may be useful techniques to control some types of weeds, though the results aren't always conclusive. Summarised crudely, raising water levels may drown (or at least significantly weaken) some emergent weeds, while lowering water levels (drawdown) may kill off submerged weeds if they dry out enough. Duration of flooding or drawdown is perhaps the most critical aspect of such methods and, if successful, re-flooding may also drown any seedlings that are triggered into growth by the disturbing processes.

Finally, it is worth considering biological senescence in the clonal sense as the ultimate, passive, long-term control for some wetland weeds. Although this type of senescence is rightly denigrated in the wider biological sense, for example when it was mooted to explain the extinction of the dinosaurs, it is a real and easily understood phenomenon when it comes to any single individual of any one species. All plants accumulate an ever-increasing load of diseases including viruses, some types of bacteria, and even fungal infections over a long period of time. This is probably the force that drives evolution to limit the lifespan of any single individual, of any particular species whether it be an animal or a plant, because the ever-increasing burden of built-in disease reduces health and vigour.

In the case of introduced weeds that have been removed from the original environment in which they have evolved, the accumulation of weakening diseases will be much slower, but it is likely that every *individual* weed carries a built-in use-by date within its genes. Single-clone weeds such as parrotfeather (*Myriophyllum*), many waterlily (*Nymphaea*) cultivars and elodea (*Elodea*) that can't produce seed without the introduction of an appropriate partner may simply drop dead of old age one day, leaving no offspring. In the case of elodea there is already evidence of this happening, as introduced populations in many parts of the world have markedly diminished in vigour over the last few decades.

Reducing light levels

Light is absorbed rapidly by water, with the red end of the spectrum more or less vanished within 1 metre of the surface, and even in the clearest waters the higher-energy blues are virtually the only colour by around 15 metres, decreasing rapidly in intensity at greater depths. Water this clear is rare in inland situations, as fresh waters are usually much more turbid or are coloured by dissolved material such as tannins, which give the characteristic dark colouration to many Australian streams. Many submerged plants are adapted to low light conditions, though to a limited degree. They can grow in situations that approach the limits of their tolerance, away from competition by more light-hungry species.

A long-term reduction in light levels can kill off such plants, although timing is likely to be critical, as species that are about to die down to some dormant part, such as a rhizome or turions, are likely to remain dormant for many months and will be unaffected by months of shading. The best time to reduce light levels to such plants is once their dormant organs have committed themselves to growth, as their stored reserves are only intended to tide the new plants through until they reach closer to the surface, where more light is available.

Reducing light by shading can be done in various ways, and even a layer of black plastic over the mud floor of a pond has been used successfully to kill off some submerged weeds, but it is not always easy to place and hold it in the desired position.

Reducing light penetration into water is also sometimes effective, and has been achieved by adding chemicals such as potassium permanganate to the water. However, most such additives are fairly expensive, and also rapidly absorbed or locked into silt and other submerged materials rich in organic matter. A simpler and more natural option is disturbing the bottom sediments, literally muddying the water. How long the effect of any such disturbance will last is partly dependent on the nature of the silt itself, as larger particles settle faster than miniscule ones, and there is also a danger of stirring up nutrients that were previously locked into the mud.

On the other hand, many Australian clays are dispersive and, once disturbed, these dissolve into fine particles that drift almost indefinitely in the slightest current – and there is always some water movement even in the smallest wetland or dam. The resulting clay-coloured water is not pretty to look at, but dispersive clays are excellent for reducing useful light levels to within just a few centimetres of the surface. When their job is done the water can be clarified by the addition of some lime, which locks the drifting particles together, into larger and heavier chunks that settle more rapidly.

Chemical control

Nearly all discussions of aquatic weed control include some details of herbicides and other chemical sprays currently approved and used for specific aquatic weeds, but it does not take much reading between the lines to see that the success rate for such programs is usually short lived, and in the long term frequently negligible. In many places where spraying programs continue they have become a perennial and sometimes escalating cost, with long-term impacts on the aquatic environment, water quality and diversity of wetland plants and animals. Many critiques of the spray-first, ask-questions-after approach have been made over the past few decades, even by some of the best-known wetland botanists. Although these have eventually resulted in increasing restriction of the types and acceptable toxicity levels of many herbicides used, it is perhaps time to query whether this simple-minded approach to wetland weed control has had any value *at all*.

One of the various reported side-effects of chemical control of weeds is fragmentation, as many of the more vigorous wetland weeds simply fall apart under stress. Not only are many of these fragments capable of growing, but they are also more likely to drift to new places where they can grow, creating new infestations. In many other cases the soil seed bank is also ignored, yet the removal of shading plants (including their own parent plants) may trigger germination of weed seeds in wetlands, because many of the most problematic species are colonisers that will always be among the fastest-growing plants in any disturbed situation.

Another problem with attempting to control wetland weeds with herbicides is the difficulty of application, especially for submerged species. Nearly all herbicides available were primarily devised for terrestrial weeds and have only been retrospectively approved for use on wetland plants as a way to widen their saleability. They continue to be used despite a marked lack of independent reviews of their use in the context of wetlands, partly because this specialised use doesn't generate enough income to be deemed worthy of detailed investigation.

Studies of the wider impacts of herbicides in wetlands are almost invariably short term, and unless there are obvious effects such as significant kills of fish or frogs, lower-level impacts such as damage to the extremely sensitive gills of aquatic animals are likely to be overlooked. Even the euphemistic language used in such studies such as 'peripheral effects on non-target organisms' seems designed to screen the reality, which is that the direct effects of these substances on diverse animal groups, the duration of such effects, and their overall impact on even the simplest aquatic ecosystems, is not known.

Even glyphosates, widely regarded as the most innocuous and rapidly broken-down herbicides, are known to be acutely toxic to many species of frogs and their tadpoles, even at the relatively low concentrations predicted for their use in wetlands. Glyphosates are particularly persistent in aquatic situations, with a half-life of five months (that is, only 50 per cent gone after that time) in water, and persisting in bottom sediments for at least a year in many cases. Nor is it necessarily the herbicides themselves that cause damage to aquatic ecosystems, as the wetting agents used to bond the active ingredient to the waxy surfaces of plants are also often suspect, and these are not necessarily included in any tests of the herbicide which is the functional ingredient.

We still know very little about the impacts of *any* herbicides and their associated ingredients in wetlands, or their long-term persistence in submerged and often anaerobic situations. It is likely that this problem will continue into the future as new products become available, only to be banned as their side-effects become increasingly known. And of course, such side-effects ripple downstream with the natural flow of the water itself, with unknown and unpredictable impacts on every wetland habitat they reach and, ultimately, affect human health.

Many chemical spray control programs continue not because they have actually achieved anything, but because councils, shires and other government agencies were long ago talked into the idea that these are the only effective, long-term control strategies for many weeds, including aquatics. In reality they are often only holding actions, carrying the assumption that this is the only cost-effective approach (as contrasted with the purportedly high cost of manual labour), even though poisoning of aquatic weeds and their associated habitats will need to be carried on forever.

The one wetland situation where chemical treatment undoubtedly offers a specifically targeted approach with minimal impacts on the wider environment is in the treatment of woody weeds, especially trees such as willows (*Salix*) and pond apple (*Annona*). Essentially, this involves cutting through the main trunk of the weed and applying the appropriate concentration of recommended herbicide directly to the fresh cut. Application methods (and their timing) vary between species and climates, and are discussed either in the individual entries where such treatments may well be justified, or in the recommended references in cases where a national plan towards eliminating such widespread weeds is already in place.

Due to the various problems and side-effects discussed above, this book makes no attempt to recommend or advise on use of potentially toxic chemicals, particularly in the hands of amateurs. Most case studies indicate that benefits are only likely to be short term except where a treatment can be specifically applied to the weed itself, and such programs often cause more long-term damage to the wider environment than benefit. In any case, the use of toxic chemicals usually requires a specific permit, and the types of chemicals that can be used (along with their legally allowed applications) are in a constant state of flux, so readers are recommended to approach the appropriate state or regional authorities for advice before considering their use.

Biological control

The first, most basic and often overlooked consideration in biological control of excessive weed growth is working out if excessive nutrient levels are contributing to the problem. If they are, finding ways to reduce nutrient inflows before they reach the water may be all that is needed to control a perceived problem. Many wetland plants are adapted to respond rapidly to unusually high nutrient levels in their soil or water, and even some normally well-behaved species may run amok and be perceived as weedy in urban situations where there is too much nutrient washing into the system.

However, it isn't always clear where fertilisers, effluent discharges and other by-products of civilisation are coming from, as they may only enter a waterway during times of peak water flow, such as the first heavy downpour after months of dry conditions. Once aquatic plants have already begun to grow excessively in such a situation control becomes more difficult, because the nutrients which fuel weedy growth are often quickly incorporated into the mud below, where the plants can easily get at them, yet disturbance or removal of the sediments is more likely to trigger algal blooms than cure the problem.

If high nutrient levels are not a contributing factor to a weed problem, it may be worth considering introduction of some animal that will feed on the problem plant as the first action, though in Australia this option is largely restricted to introduction of various insect and disease biocontrols. Unlike in much of the rest of the world, there are very few native herbivorous animals in Australia that are likely to have any noticeable impact on fast-growing aquatics, and our indigenous freshwater fishes are nearly all carnivores descended from primarily marine families. A much greater range of herbivorous fishes is available elsewhere, of which grass carp (*Ctenopharyngodon idella*) seems to be regarded as a relatively innocuous species, and it has been widely stocked to reduce excessive growth of submerged plants with unpredictable degrees of success. In New Zealand the results have not been spectacular, and as the grass carp feeds indiscriminately on a wide range of plants it is as likely to eliminate indigenous plants as weeds.

Domestic ducks have also been used overseas to control excessive growth of some aquatic plants including duckweeds, *Potamogeton*, *Ruppia*, *Chara*, *Sagittaria* and *Vallisneria*, not all of which are regarded as weeds in Australia. Wild ducks (particularly black ducks, *Anas superciliosa*) will do a comparable job if encouraged to congregate in an area by feeding them regularly. Bird numbers must be maintained for some time to achieve reasonable results, usually several years, or the problem is likely to return. However, water quality is also likely to deteriorate if unnaturally high duck populations produce excessive droppings in proportion to the volume of the water body.

The recommended stocking rates for domestic ducks as a permanent control for submerged weeds are from 20 to 40 per hectare, depending in part on the size of the breed used. Where unwanted vegetation is growing deeper than 60 cm, domestic geese have been used instead, but they must be penned to confine them to the bank and water body or they will graze on terrestrial grasses in preference. However, waterfowl will not feed on some weedy aquatics such as *Pistia* and *Eichhornia*.

The primary arena where biological control is most likely to make a significant impact on weeds is not in private hands, but is rightly a monopoly of federal government: the introduction of biological control agents. Plants introduced outside their natural range are often weedy because they have been freed of the natural constraints on growth they have co-evolved with, from diseases to herbivores that specialise in feeding on various stages of their life cycle ranging from newly formed seeds to soft stems, dormant tissues and young leaves according to their predilections. Introducing the most effective species of these controlling organisms is obviously a good idea, but it can backfire, especially if there are also native plants that could be attacked or damaged by new and unfamiliar insects, disease or other biocontrol agents.

Successful biological control is a wonderful and elegant thing, and already some weeds such as salvinia (*Salvinia*) and water hyacinth (*Eichhornia*) have been markedly affected by just a few introduced insects, the latter most dramatically in some overseas examples rather than in Australia; specialised weevils are among the most successful groups of biocontrol agents. The main brake on introduction of potential biocontrol species is their possible side-effects, which need to be researched in some detail before any new species is considered for release. We don't need any more national disasters such as the cane toad (*Bufo marinus*), originally introduced to control insects in sugarcane crops, but now arguably the most invasive plague species in otherwise relatively undisturbed wetlands.

Competitive planting

Another form of biological control of wetland weeds is planting bare areas or newly created wetlands with more desirable plant species, and this should be done as soon as possible, rather than allowing random weeds to appear from wind-blown seed or carried in by visiting waterfowl. This inevitably restricts the types of planting to be done to species that can tolerate relatively raw and unnatural submerged soils, as has been discussed in the second edition of *Planting Wetlands and Dams* (Romanowski 2009). The difference an early planting makes to species profiles and overall diversity can be dramatic, as can be seen in two constructed urban water treatment wetlands at Rushworth, in north-central Victoria.

The first of these was constructed five years before I was asked to plant it, and had already been colonised by all three species of *Typha* – both native and naturalised in Australia (see plate 2). Despite an expensive attempt to remove most of these weeds with a long-reach excavator before replanting, the cumbungi has repeatedly returned and remains by far the most abundant plant, except in the deeper, open water sections where nothing grows. By contrast, in a later wetland constructed 2 kilometres away which I planted within a few weeks of its completion, around eight indigenous species make up around 95 per cent of the plant mass, and not a single *Typha* seedling has appeared, as the seed is so small it is easily smothered or shaded out by virtually any other wetland plant.

Competitive plantings don't necessarily just shade out seed of later arriving plants. Many aquatics are known to be allelopathic (or antagonistic), using some interesting arsenals of toxic chemicals against other plants. These range from simple organic acids to more complex poisons such as alcohols, tannins, alkaloids, sulfides and even long-chain fatty acids. Most wetland plants probably only have a limited arsenal of up to around half a dozen chemicals they release into water, or into soil through their roots, to deter other species, while others vary in their abilities from being capable of just producing one or two types of chemical to ten or more different types of toxin to discourage their competition.

As most of the study in this area has been done overseas, not much is known about which Australian plants use allelochemicals, though it is often easy to guess. Some *Eleocharis* species strongly discourage competing plants, which is probably why they are often found in dense, single-species stands. Several exotic *Myriophyllum* have also been shown to effectively reduce growth in a number of other plants with the help of cyanogenic compounds (among others), and it is likely that at least some of the larger Australian species do so as well.

Many other Australian aquatics must also be toxic if their non-native relatives are any guide, ranging from *Brasenia*, *Ceratophyllum*, *Hydrilla* and at least some *Juncus*, to *Nymphaea*, *Potamogeton* and *Vallisneria*. The native *Bacopa monniera* produces only nicotine as a deterrent, while an American relative of the native seagrass *Posidonia* is known to use around a dozen complex chemicals in its quest to exclude competitors, which it does very effectively as anyone who has swum over enormous, single-species beds of this plant can testify. There are undoubtedly many more species to be added to this brief list and, without even studying the compounds they produce, allelopathy can often be confirmed simply by growing suspected species with a range of possible competitors, although not all of the toxins any particular plant can produce will necessarily be manufactured at all times.

The most straightforward form of competitive planting is replacing indigenous vegetation that has been removed from stream banks and other situations. This isn't always possible immediately, because we have also created many other problems in our rush to replace native riparian plants, ranging from steeply eroded and flood-prone slopes to the difficulties of establishing anything under a canopy of introduced weeds such as willows, blackberries and lantana. It may be necessary to initially shade out the weeds themselves by planting evergreen trees such as blackwood wattles (*Acacia melanoxylon*) with a dense leaf canopy further from the water, eventually removing and replacing them with lower-growing wetland plants once the shoreline has stabilised and the competing vegetation is manageable.

When all else fails

Early in the 21st century we are still at a stage where we may be able to eliminate many wetland weeds, or at least effectively contain their further spread, in an attempt to retain the diversity of native wetlands that have already endured so much change over the past two centuries. Nor is it too late to curb further escapes from gardens and aquariums, some of which are still being legally traded (even though their invasive potential has already been well documented), while many other potentially weedy species should simply be prohibited on the precautionary principle.

However, in many other cases the battle has been lost long ago, and failing the discovery of effective biocontrols, or changing the ways in which we use and move water around to reduce the impacts and further spread of some entrenched weeds, we are going to have to learn to live with many of these problem species. The outlook is not always dire: for example watercress (*Nasturtium*; see plate 27) is probably as widely distributed as it could be, yet little in the way of adverse effects has been documented for this sometimes abundant plant. Similarly, at least some of the introduced waterlilies (*Nymphaea*) are ecologically comparable to the native species of northern Australia, although much could still be done to reduce their further potential spread by seed.

Many other weeds are relatively specialised colonisers, growing most vigorously on disturbed ground including rice fields, so that more appropriate crop rotations and management should be enough to minimise their impacts. Part of what we need to be doing now is assessing which weeds we can live with, whether their removal is likely to make a real difference to the ecological health of natural waterways, and make more of a study of which weeds may actually create new types of habitat, or may be an acceptable replacement for things we have lost.

The reality is that most wetland weeds will never be completely eliminated, and many will successfully integrate into the Australian environment over time. In some far-future world these will eventually evolve into distinctive Australian forms and ultimately species, adapted to the conditions of this country in their own right, with their own suites of disease organisms, competitors and other controlling agents. But by then they will no longer be weeds.

Weedscapes. An inland backwater (above) completely dominated by weeds including willows, cat-tail and Mexican waterlily – the only indigenous plant here is ferny azolla growing as a red, floating carpet, peppered with the white flowers of cabomba. At centre a dense, floating mat of alligator weed in an urban waterway crowds out even vigorous native sedges. Below, arum lily has invaded an ephemeral paperbark swamp in Western Australia; the dense cover of duckweeds suggests both the sheltered nature of this wetland, and also high dissolved nutrient levels.

The created wetland above was left unplanted for several years, as a result of which it is now dominated by cumbungis, among the most invasive indigenous wetland plants. In the foreground the relatively small narrowleaf cumbungi contrasts with a faster-spreading stand of broadleaf cumbungi in the distance. At left below a closer view of narrowleaf cumbungi in flower, with broadleaf cumbungi at right, showing the typical gap between the upper, male flowerhead and the female flowerhead below.

By contrast, there is not a single cumbungi plant in another created wetland (above) less than 2 km from the one opposite, because the tiny seed was unable to germinate among the diverse indigenous species that were planted here within weeks of the wetland filling. At centre and below, the introduced cat-tail (closely related to cumbungi) is spreading rapidly in south-eastern Australia, even into apparently pristine waters such as the headwaters of this natural lake formed by a landslide in a national park. Cat-tail can usually be told apart from cumbungis by the bluer leaves, the much darker female flowerheads, and the smaller gap between male and female heads.

Three opportunistic annual grasses that colonise disturbed habitats: at top barnyard grass, with annual beardgrass at centre. Even the indigenous common blown grass (below) has been regarded as a weed in recent drought years, when it has become one of the dominant plants of drying lake floors in south-eastern Australia.

Serious potential wetland weeds of the future: the pampas grass plants above left are seedlings just several years old, while reed sweetgrass (above right and below) grows so vigorously even in slow-moving water that it can choke the flow of streams.

Water couch (at top) was regarded as indigenous until recent decades, and may well be as there is no real evidence to confirm that it has been introduced; its management hinges on whether it is perceived as a vigorous introduced weed, or a minor habitat plant. By contrast, reed canary-grass (left) and Parramatta grass (above) are relatively worthless, introduced pasture grasses that were never much of a success in their original roles, but have become widespread weeds in disturbed wetland situations.

All true reeds are large grasses, but while the introduced giant reed (above at left) is an increasingly common weed across much of southern Australia, the native common reed (below) is a significant habitat plant. Other so-called reeds such as branching burr-reed (above at right) are not grasses, though they belong to related families.

Sedges are closely related to the grass family, and include many species that are regarded as weeds in various contexts. Drain sedge (above at left) is the most widespread and abundant introduced sedge, though it rarely invades undisturbed wetlands. The native tall spikerush (below) is still sometimes planted in created wetlands even though it can form dense, monospecific stands in water up to 3 m deep. Budding clubrush (above at right) was still being sold by some nurseries until recent years, allowing it to extend its range.

The nursery trade continues to sell weedy sedges including many *Carex* species. The blue plant above at left is *C. flacca*, an introduced weed collected in Tasmania, which is still being widely offered as variants of several native species. Many New Zealand *Carex* species are regarded as weedy even in their native land, and are already proven weeds in Tasmania, yet are still freely available on the mainland. Makura (below) has the potential to radically change the appearance and habitat values of many peaty wetlands; at top right, a coloured form of *C. testacea* in a garden centre.

The aptly-named spiny rush (below and above at left) is one of the most unpleasant wetland weeds, growing on a range of soils, tolerating salt and drought, and is difficult to deal with once an infestation is well-established. Other introduced rushes such as jointed rush (above at right) may be less conspicuous but are equally widespread.

Toad rush (above) is an annual species of uncertain origins, but like the introduced small-headed rush (at left below) is relatively easily shaded out by taller wetland plants. Soft rush (at right below, cultivar 'Spiralis' at a garden centre) is an established wetland weed in some places, with considerable potential to spread further from named clones that have apparently been illegally introduced in recent years.

'Indigenous' weeds. Rainbow nardoo is a valuable habitat plant in eastern Australia, but an unwelcome weed in Tasmania after its deliberate introduction into a stream south of Hobart. Floating bladderwort (at left below) is a so-far minor weed spread by the water garden trade, while several paperbarks including *Melaleuca parvistaminea* (at right below) have naturalised as a result of planting outside their natural range.

Natives with an unusually wide range. Waterbutton (at top) has been treated as a South African weed by some sources, despite evidence of its pollen in ancient lake bed sediments published decades ago, while the pollen of purple loosestrife (at left) was found in deep sediments more recently. By contrast, no one has ever doubted that hornwort (above) is indigenous, even though this cosmopolitan plant has made its way to such remote locations as Pacific islands, by unknown means.

It is not clear whether some wetland plants such as *Persicaria lapathifolia* (above), *Berula erecta* (at left below) and *Rorippa palustris* (at right) are indigenous or introduced. Ideally, genetic testing of a number of Australian populations should be carried out against populations from other countries within their native range, with the aim of resolving whether any of the ten or so species in question are weeds.

Fairy ferns are among the most widespread indigenous floating plants, and can become weedy in high nutrient situations. Pacific azolla (above) only turns pink-red when stressed or in strong sunlight, while ferny azolla (below) is red at all times. At centre, the introduced weedy Toowoomba canary-grass on the shoreline of a farm dam smothered by Pacific azolla. The central photograph was taken on the first hot day of summer; three weeks later there was no trace of the azolla, but the resulting decay had already lowered dissolved oxygen levels dramatically.

Most duckweeds in Australia are indigenous, including this mix of the tiny *Wolffia australiana* with *Lemna disperma* (above), also growing with a few tiny plantlets of Pacific azolla. Dense growth of duckweeds such as in the created wetland at centre is an infallible indication of high nutrient levels, in this case from urban runoff. Water lettuce (below) is indigenous in the Northern Territory, but has also been introduced into the eastern states where it is a potentially serious weed.

Water hyacinth (above, smothering a slow-moving stream, and flowering at left below) has been widely spread through tropical regions as an ornamental plant. Once regarded as one of the worst weeds in the world, biological control is making significant inroads into feral populations in many places, as is also the case for the floating fern salvinia (below at right).

Landscape or weedscape? The attractive riverside scene above is dominated by an unknown introduced waterlily that has increased its surface coverage tenfold in the past 15 years, yet is not as weedy as the tropical Cape waterlily (centre), or the more temperate-growing Mexican waterlily below.

Other plants with floating leaves that are still available from nurseries include spatterdock (above), a close relative of waterlilies that will grow vigorously in even deeper and cooler waters. Water hawthorn (centre and below) is a relatively minor South African weed at present, but has the potential to spread further with global warming. The dense band of aquatic vegetation along the far side of the stream at centre is the introduced aquatic weed, parrotfeather.

Three potential weeds available through water garden nurseries. Water poppy (above) and the pink-flowered *Rotala* (centre) are already patchily naturalised as far south as Victoria, and could expand in range with global warming, while Chilean rhubarb (below) is a specialised streamside herb that would be relatively easy to control if it became feral.

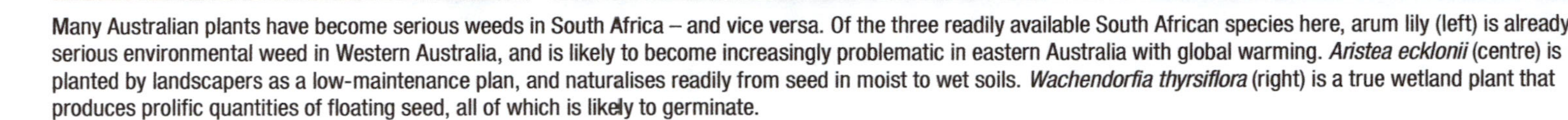

Many Australian plants have become serious weeds in South Africa – and vice versa. Of the three readily available South African species here, arum lily (left) is already a serious environmental weed in Western Australia, and is likely to become increasingly problematic in eastern Australia with global warming. *Aristea ecklonii* (centre) is planted by landscapers as a low-maintenance plan, and naturalises readily from seed in moist to wet soils. *Wachendorfia thyrsiflora* (right) is a true wetland plant that produces prolific quantities of floating seed, all of which is likely to germinate.

A number of very weedy aquarium plants appear to have been deliberately naturalised for wild harvesting purposes, an environmentally disastrous move that was also self-defeating as cabomba (above) and Senegal teaplant (at left below) are now prohibited from sale in most parts of Australia. Parrotfeather (below at right) grows so profusely in nutrient-rich situations that it most likely naturalised from garden refuse, but as there only appears to be a single, female clone in this country it can only spread from cuttings.

Water thyme (above) is an indigenous submerged plant that grows rampantly in warmer climates, but is also a significant habitat plant. Its introduced relatives Canadian pondweed (inset, also below at right in an irrigation channel with barnyard grass) and dense waterweed (below at left) are comparably weedy in a wider range of climates, but as these are single-sex clones they are unlikely to have a significant long-term future as weeds.

Streamside weeds: although not wetland plants, both lantana (above) and blackberries (centre and below) often grow as dense thickets along rivers, impeding access and providing habitat for vermin rather than indigenous species. There are well-defined programs providing advice on control of both of these nationally significant weeds, both on the internet and as informative hard copy manuals.

Riparian weeds: willows (centre and above) are worthless as habitat for virtually all indigenous animals, dump semi-toxic leaves into rivers at the end of their growing season, contribute to the collapse of river banks during periods of flood, and transpire immense volumes of water during hot weather, significantly reducing environmental flows. Although the related poplars (below) are not as common or widespread, this dense stand surrounding a small wetland has probably tripled the natural evaporation rate over the course of a hot summer, compared to a similar-sized but poplar-free pool nearby.

These three invasive, tropical weeds are still relatively restricted in range so there may be some prospect of controlling them, but once established they can radically change the nature of wetlands. Pond apple (above) has spread through thousands of hectares in national parks and in the fresher areas of mangrove swamps, olive hymenachne (at left below) is a pasture grass deliberately released as recently as 1988 and probably capable of spreading into undisturbed wetlands, while mimosa (below at right) forms dense, thorny stands shunned by both humans and most animals.

Weeds of cooler waters and places include watercress (above) which has probably already occupied much of its potential range in Australia, and the less aquatic and also less weedy blue water-speedwell (centre). Common starwort (below) is seen here growing abundantly in a stream that had dried out completely in the course of a dry summer the year before summer, but was mostly gone a few months later.

Some wetland weeds grow in shallow water, but form floating, attached mats that trail out over deeper waters as well. Alligator weed (above) has been deliberately spread around urban areas as an edible plant over the past decade. Swamp ludwigia (centre) is already naturalised in many places, and has the potential to spread much further if it continues to be legally available from nurseries. Water primrose (below) has usually been regarded as an indigenous plant, but may be a post-European arrival.

Willowherbs are closely related to *Ludwigia* species. Though some indigenous species such as hairy willowherb (above) spread freely in disturbed, seasonally wet situations, they are not significant weeds. By contrast, the introduced tall willowherb (at right below) was spreading rapidly in the Geelong area until it was systematically eradicated, while pinkweed (below left) is a common but short-lived weed in nurseries and drains.

Several native and introduced species of the water plantain family are regarded as rice field weeds. A few plants of water plantain (at left below) and giant arrowhead (below at right, growing with water couch) have established themselves in a corner of the otherwise well-cultivated crop seen above. Note the floating mat of cyanobacteria (sometimes referred to as blue-green algae) pushed into the channel foreground by a strong wind; this would normally have been spread so thinly that it would be inconspicuous.

Freshwater algae grow as mats, threads, in small floating colonies or even as single cells, but feathery stonewort (above, and in the inset covered with a simpler algal species) and its relatives look superficially like some flowering plants. Most smaller species of algae are only conspicuous in high nutrient situations such as in the creek at left below, where the algal mats are dramatically tinted red by bacterial growth, or in farm dams where nutrients are constantly stirred up by cattle as at below at right.

Seaweeds are more complex marine algae, and the invasive Japanese kelp (above, and at centre as the dominant plant in an enclosed harbour) is a fast-growing annual easily recognised by its prominent midrib; the convoluted sporophylls at the base are only obvious on mature plants. Although *Caulerpa taxifolia* (below) is becoming a major ecological disaster in the Mediterranean and elsewhere, in Australia decisions on appropriate control of this species are complicated by the presence of indigenous populations at least as far south as the Brisbane area.

3

Native plants as weeds

Before discussing the conditions under which diverse Australian plants have come to be considered as weeds, it is worth defining several terms used throughout this book in varied contexts. The term **native** is a general category covering any plant that was already in Australia before European settlement in 1788, even if it may have been introduced not all that much earlier from south-east Asia or by earlier European voyagers. It covers a multitude of species across the whole continent, but although a grass or sedge from Western Australia may be native in that state, this doesn't mean it is necessarily native in eastern Australia as well.

Plants that are native in your local area are **indigenous**, truly native in every sense of the word, though they may also be much more wide-ranging. Although such plants may be regarded as a single species across much of this country at the present time, they may vary considerably from one place to another, and in some cases may ultimately be separated into several similar-looking species once their biology is better understood. For the purposes of this book, **exotic** means much the same thing as **introduced** – in most cases brought in from other countries, though any native plant that has been moved outside its natural range by humans should also be considered an introduced weed.

Before considering diverse aspects of native wetland plants as weeds, it is worth commenting that many indigenous species are included in lists and manuals of weeds simply because they may sometimes obstruct flow in open water channels and canals. Some of these are discussed briefly under the appropriate species entries later in this chapter, or in the main compendium of weeds if they are related to more serious weeds, and may be confused with them in some situations.

However, most such species are not included because the channels are the problem, not the plants. In a country where water losses from evaporation and seepage through the walls of earthen channels have been reported at 80 per cent and higher, peaking during the hottest times of year when the water is most in demand, it is obvious that the use of open channels is an inappropriate design (see plates 23 and 25). Apart from direct water losses, there is also the issue of increasing concentration of salts: if evaporation was the sole cause of loss, this could mean a five-fold increase in salinity by the time water has been delivered to its destination. Nor is it just weeds that colonise open channels, as these are also conduits for the spread of diverse introduced fishes, disease-carrying snails and other aquatic vermin that are likely to find a wider range of climates and habitats available to them as global warming progresses.

Fresh water should be distributed through pipelines rather than an open grid of channels, as pipes are more easily cleared of such unwelcome passengers at the entry point, while any young fry that slip through preliminary filters are likely to be too small to survive a journey of any length through pitch darkness, without the chance to feed. The cost of piping water on such a scale may seem prohibitive if you look at the volume flowing through some irrigation systems, but keep in mind that much of this water is wasted. A pipeline only needs to carry a fraction of the water moving through an open channel, and the cost would be trivial compared to public monies spent on projects such as freeways, which are far less significant to our long-term survival and sustainability on this driest of all livable continents.

Native or introduced?

For the past two centuries many aquatic plants found overseas as well as in Australia have been regarded as indigenous, unless there was good reason to think they had been brought in by European settlers, either deliberately or by accident. However, more recently a few of these species have become weeds in the eyes of a relatively small but apparently influential group of botanists, and have increasingly been treated as such in the Floras of the past two decades.

The starting-point for this turnabout was an ambiguous paper (Kloot 1984) that considered nearly 100 plants which had generally been regarded as being indigenous in Australia, but which in the author's opinion were actually introduced within the boundaries of South Australia. Although Kloot tempered his opinions with the qualifier '... assuming that the taxonomic determinations are correct', it was clear that his often controversial, and now mostly obsolete, suggestions were made on the basis that if there seemed to be too large a gap between their Australian range and the places they were found overseas, they must be aliens.

Slightly more than half of the species included on this hit list happen to be wetland plants, which is not surprising as it is well known that many aquatic and wetland plants are very wide-ranging, even though we still have little idea of how many of them spread over great distances. Consider the relatively specialised aquatic water shield (*Brasenia schreberi*), found from Africa to the Americas, and through much of Asia southwards to northern Victoria. No one doubts that this plant is indigenous throughout this immense range, or has even tried to formally separate the many and distinctive regional forms that are known to exist worldwide. Nor does anyone know how it has managed to spread so far, given that its relatively heavy seeds (which aren't produced all that often) can not blow with the wind, and are unlikely to survive long in seawater.

The most likely explanation for this wide range is that the seed of this species is adapted to being carried by birds, presumably surviving undigested in their gut while they travel in search of new homes. The only reason water shield is unanimously accepted as an accomplished traveller rather than an invader from far-distant lands is that there are no significant gaps (disjunctions) in its range, but there are many other indigenous wetland plants that are also well adapted to skip even over oceans. These have been discussed in my earlier book *Planting Wetlands and Dams*, and need not be considered here in any detail.

It is worth considering how Kloot's opinions on the 50 supposedly introduced wetland plants have fared in the eyes of others, since their first airing as putative weeds a quarter of a century ago. Of the plants he says are '... believed, at least by early botanists, to predate European settlement' (p. 63), most are *still* universally accepted as natives because there has never been any evidence to the contrary. Before looking at the very few species on it that are still regarded as possible introductions, let us list the 80 per cent agreed to be native using their currently accepted names, with the names they were known by in 1984 in brackets where appropriate.

These are *Bolboschoenus caldwellii* (as *Scirpus maritimus*), *Bolboschoenus medianus* (as *Scirpus fluviatilis*, now *B. fluviatilis*; this is actually a different species not found in South Australia, but native in the eastern states), *Calystegia sepium*, *Centella cordifolia* (as *C. asiatica*), *Centipeda minima*, *Ceratophyllum demersum*, *Cladium procerum* (as *C. mariscus*), *Cotula vulgaris*, *Gonocarpus micranthus* (as *Haloragis micrantha*), *Gratiola peruviana*, *Halophila australis* (as *H. ovalis*), *Hydrilla verticillata*, *Isolepis cernua* (as *Scirpus cernuus*), *Isolepis fluitans* (as *Scirpus fluitans*), *Isolepis inundata* (as *Scirpus inundatus*), *Isolepis marginata* (as *Scirpus antarcticus*), *Juncus planifolius*, *Lemna trisulca*, *Lobelia anceps* (as *L. alata*), *Lythrum salicaria*, *Montia fontana*, *Myriophyllum* (three endemic species lumped together as *M. elatinoides*, an exotic species), *Phragmites australis*, *Polygonum plebeium*, *Potamogeton crispus*, *Ruppia maritima*, *Schoenoplectus litoralis* (as *Scirpus litoralis*), *Schoenoplectus pungens* (as *Scirpus americanus*), *Schoenoplectus tabernaemontani* (as *S. lacustris*), *Selliera radicans*, *Spiranthes australis* (as *S. sinensis*), *Sporobolus virginicus*, *Stuckenia pectinata* (as *Potamogeton pectinatus*), *Triglochin striata*, *Typha domingensis*, *Vallisneria americana* (as *V. spiralis*) and *Zannichellia palustris*.

Despite some name changes, in *all* of the above cases what we know about these species and their ranges is largely unchanged, so they are still regarded as indigenous wetland plants that have a wide natural range. A few species on Kloot's list may well be introduced weeds though this remains unproven, while the status of others remains in limbo. Purple loosestrife (*Lythrum salicaria*; see plate 13) is now largely accepted as indigenous after lake-floor core samples from long before European settlement found pollen of either this species or its close relative *L. hyssopifolia*, which has also been regarded as a possible introduction. The remainder of Kloot's list of doubtful natives is short: *Batrachium trichophyllum*, *Berula erecta*, *Ludwigia peploides*, *Paspalum distichum*, *Persicaria lapathifolia* and *P. hydropiper*, and *Rorippa palustris*.

Of these *Berula erecta* (see plate 14) and *Persicaria hydropiper* are only regarded as minor weeds at best, showing little sign of spreading, and given that they may be indigenous these species are not considered further. *Batrachium trichophyllum* is a peculiar candidate for weed status: though it is widespread in south-eastern Australia, it is uncommon except in some higher and cooler areas remote from older settlements, and still more uncommon near the major urban areas where it should have been most abundant if it was a recent arrival. This is an unusual distribution pattern fitting far better with the notion that *Batrachium* is a not particularly successful immigrant from some earlier time.

By contrast, both *Persicaria lapathifolia* and *Rorippa palustris* (see plate 14) are either seriously invasive, short-lived colonisers that can completely take over large areas of drying wetlands and river banks, or if indigenous they are desirable, short-lived colonisers that bind recently exposed wetland soils against erosion during periods of drought, and provide much-needed shelter for a variety of wetland animals. You can take your pick of these choices, because I don't believe there is enough evidence to decide one way or the other. As there is not the remotest prospect of controlling, let alone eradicating, either of these species they are also not considered further.

Ludwigia peploides (see plate 28) is discussed with its various introduced relatives, and *Paspalum distichum* (see plate 6) is discussed with other grassy weeds simply because it grows too enthusiastically in high nutrient situations. As Kloot's list was confined to the geographic boundaries of South Australia, the status of other weedy species included in this book such as water plantain (*Alisma plantago-aquatica*; see plate 30) weren't questioned in his article. However, this species is mentioned here as another far-ranging plant still universally accepted as indigenous, though its status may warrant further investigation.

It should be clear that Kloot's list is virtually obsolete when it comes to wetland species, and that his inferences are no basis for weed management in Australian wetlands. This can be seen most clearly in the case of one widespread and abundant 'weed' on his list which has yet to be mentioned, declared to be an introduced species not all that long ago. We have enough monumentally serious problems (not just weeds) to deal with in the next few decades as fossil fuels run dry, without wasting time, money and effort on trying to manage plants that were never weeds in the first place, so it is time to spell out a cautionary tale.

Waterbutton: a cautionary tale

Waterbutton (*Cotula coronopifolia*; see plate 13) is a succulent-looking daisy with fleshy, divided leaves, and with its masses of vivid-yellow button flowers is a conspicuous plant in many saline and freshwater wetlands across much of the southern half of Australia. This long-flowering and distinctive plant was accepted as indigenous until the 1990s, after which time it was treated as a South African weed in various Floras because of its inclusion in Kloot's lists. It is also the best-documented example of how poorly supported a charge of exotic origin can be.

Waterbutton was a familiar plant around the Port Jackson settlement where it was first collected by Robert Brown, just after the turn of the 18th century. He had already seen the same species in southern Africa on the journey here (it is also native in southern South America, and possibly in New Zealand), so he included it on a list of 29 introduced and *possibly* introduced species for the Port Jackson area. Of these, waterbutton was one of two species tagged with a question mark, because of its wide distribution around Port Jackson which suggested it may have been long established there.

As exploration of Australia continued over the next half century, waterbutton was found growing in wetlands both inland and along most of our southern coasts. If it *had* been a recent introduction at Port Jackson, it must have been one of the fastest spreading weeds on record, yet it has not been observed to spread into any new area since those early records were made. In comparable areas of the northern hemisphere where it is most certainly introduced, waterbutton has spread at a much slower rate, achieving its present range over centuries.

A distribution covering all southern hemisphere continents is not unusual; for example, the sedge *Ficinia nodosa* (which no one doubts is native) has a similar range, although it is generally found on higher and drier land around coastal and some inland marshes, rather than actually in them. Other native saltmarsh species are more restricted in range, for example *Samolus repens* and *Selliera radicans* that are 'only' found from Australia to South America, and *Juncus kraussii* which is abundant in both Australia and South Africa, though it may be a different subspecies there.

If waterbutton is native to all these continents, separated by the two widest oceans on Earth, or even just native to South Africa and South America but not Australia, how did it spread? As this species will tolerate salinities nearly twice that of seawater, and broken-off pieces I have floated in seawater have started putting out roots after four months, it isn't hard to imagine how living pieces of this plant could survive a sea journey. Stranger and much more fragile things have floated across the Indian Ocean to Australia, not least two sub-fossil *Aepyornis* eggs from Madagascar (the largest of all bird eggs known), which were still intact when they washed up in Western Australia. Broken-off pieces of waterbutton are far less delicate and vastly more common, as can be seen along some of our southern coasts after a south-westerly storm.

If waterbutton was in Australia before European settlement, its pollen should have shown up in some of the many sediment cores taken from lakes and swamps in south-eastern Australia over the past few decades – assuming that anyone was looking for it. Most pollen studies of this kind primarily concentrate on terrestrial pollen to gain an understanding of the forests, grasslands and associated vegetation of pre-European Australia. These forests ran for many kilometres in all directions, and the pollen of eucalypts, grasses and even sometimes terrestrial sedges is much more abundant than that of the aquatics, which can only grow as a thin band around the fringes of deeper lake waters.

Pollen records of aquatic plant species are incidental to studies of terrestrial pollen, and palaeobotanists generally don't bother putting much work into plant groups which aren't central to their main interests, because these don't tell you about the wider terrestrial communities which are the central focus of most studies. This is why the species-rich daisy family (Asteraceae) to which waterbutton belongs is almost invariably just reduced to the two main subfamilies for which pollen can be quickly identified, the Tubuliflorae and the Liguliflorae.

In most articles looking at pre-European pollen buried in Australian lake beds, no effort is made to separate these into the hundreds of species potentially included. However, three papers published decades ago by John Dodson (1979, 1986 and with Wilson in 1975) *did* differentiate *Cotula* pollen from the rest of the Asteraceae. Now a professor of Earth and Geological sciences in Western Australia, Dodson's 1983 study of modern pollen rain in New South Wales forms an important underpinning of the major synthesis of south-eastern Australian core samples by D'Costa and Kershaw (1997).

In many cases identification of fossil pollen to species level is difficult or even impossible depending on the preservation of the granules, so Dodson's papers don't specify which of the three wetland *Cotula* found in lowland regions it comes from. These are waterbutton itself, plus the uncommon *C. vulgaris* (an indigenous variant of a species also found in South Africa) and *C. australis* (also regarded without evidence as possibly introduced). A fourth native species from the high country (*C. alpina*) is found in moist, alpine herbfields or even *Sphagnum* moss beds, rather than near open water, and is an improbable candidate.

However, other records in the same papers provide an unambiguous comparison in the form of *Typha* pollen, as only one of these four *Cotula* species produces a similar volume of pollen to the prolific, wind-pollinated cumbungis, and that is waterbutton. An informal comparison I carried out for the three most plausible *Cotula* candidates showed that waterbutton produced on the order of 1000 times more pollen over its six month (and sometimes longer) flowering season than the other two did together. Furthermore, the others only flower when growing on moist to wet soil, so nearly all of their pollen fell onto the soil itself, and not onto the water's surface as was the case for waterbutton.

There are two interesting points to be drawn from Dodson's papers, the first of which is that in two of them *Cotula* pollen only goes back to around 5000 or 6000 years before the present (BP). In the third it appears earlier around 10 000 years BP, but is missing from the 5000 years also sampled before that. If this is not just a statistical fluke, waterbutton may only have been here for a few millennia. This could also explain why it has not yet had the chance to develop the small differences *C. vulgaris* has from its South African cousin, which are the reason this less abundant plant has never been accused of being a recently introduced weed.

The other aspect is more disturbing. Despite published evidence that waterbutton was probably widely distributed here even in the earliest days of European settlement, that broken-off pieces can survive ocean crossings better than fragile giant eggs (seeds carried by migratory birds are another possible means of spread which hasn't been considered here), despite the complete absence of evidence that it has been introduced by human agency, and despite the presence of its pollen in lake core samples published 30 years ago, it was first declared a weed in the third volume of the *Flora of New South Wales*, to be followed shortly after in the *Flora of Victoria*.

Waterbutton is not the only indigenous plant that has come perilously close to being treated as a serious, introduced weed. Purple loosestrife (*Lythrum salicaria*) was also hanging by a slender thread for a while, until pre-European pollen samples were recovered from Wingecarribee Swamp in New South Wales. As with waterbutton, there is an ambiguity here as another species (the annual *L. hyssopifolia*), which is also regarded as 'possibly' introduced could be involved, though this annual grows primarily on seasonally damp ground rather than in water.

In another sense, it doesn't matter all that much – if either one of these two species (which have a similar natural range) has managed to arrive in Australia under its own steam, so also could the other. In our present state of ignorance we should be looking more closely at the genetic make-up of the very few wetland plant species still living under the threat of potential weed status. The technologies are already there and are becoming increasingly affordable as they are refined, so it would not be a great expense to clear the air on this issue once and for all. A complicating factor is the compromise position that has been used to slip around this problem in more recent years – the admission that there may well be 'native' strains of a particular plant, yet implying that such species as toad rush (*Juncus bufonius*; see plate 11) may include introduced forms.

It is likely that waterbutton has become more abundant since European settlement as it is not eaten by livestock. However, there is little evidence that it has changed in overall distribution as suggested by Randall (2007), as the earliest records by those few Europeans with an interest in indigenous plants show that it was already well established in most climates that suited it. Even outside the heavily grazed paddocks where it thrives particularly, it may often be a significant habitat plant. For tadpoles and a fair few types of invertebrates in shallow waters, the mild toxicity of waterbutton foliage is irrelevant, and some of these smaller animals find it a valuable shelter species, if not a source of food.

Native plants on the move

The potential problems which could be caused by plants found in one part of Australia being introduced elsewhere should not be underrated, and will become an increasing problem with the combined popularity of 'native' gardening and climate change unless steps are taken now to minimise future effects. The problem is not confined to wetland plants alone, and there are many people who are concerned that many fire-adapted terrestrial plants grown in gardens have the potential to become ecological invaders after wildfire.

The situation is perhaps less drastic for most indigenous wetland plants, but already some suspiciously mobile species have been noted in recent decades, and these are discussed later in this chapter along with plants that are only regarded as weedy in rice fields and drains. Other indigenous wetland plants more widely regarded as weeds, or which are closely related to some introduced weeds, are discussed in the next section of the book. A complete ban on sale or movement of so-called native plants would be impossible to enforce, especially as many of these plants have already been widely distributed by the nursery trade, not just by water garden nurseries.

However, in my experience the great majority of people who run nurseries would voluntarily restrict trade in suspect or potentially weedy species if they were told, in plain English rather than bureaucratic waffle, why this was desirable. A register of some native species that should not be traded beyond certain areas would go a long way towards preventing the spread of new weeds (both terrestrial and aquatic), and would also be welcomed by most prospective purchasers. Ideally, such a list should also suggest closely related species that are actually indigenous in certain areas, as in most parts of Australia there are equally attractive, similar-looking and more ecologically desirable species available. The only reason such plants aren't offered by nurseries is simply because they don't realise there is no reason to import potential native weeds from elsewhere.

Minor indigenous wetland weeds

Acaena (Rosaceae)

Bidgee-widgee or biddy-biddy (*A. novae-zelandiae*) is a low-growing, running herb which may cover large areas of damp ground, forming large seedballs that hook readily into fur, down feathers and clothing. As a result it has become unnaturally abundant in some places, carried by various domestic animals including horses, dogs and sheep, and becomes a dominant plant in damp pastures where more desirable pasture species are overgrazed. Massed seedballs may also be spread from such pastures by grey kangaroos.

Uses: a minor habitat plant.

Preferred growth conditions: not a true wetland plant, but found in a wide range of conditions from the fringes of wetlands, to better-drained situations in higher-rainfall areas.

Confusing species: there are several other species in this genus, and at least one uncommon natural hybrid, but none of these appear to be weedy in wetland situations.

Environmental impacts and values: a minor weed in some places, otherwise a widespread native plant although little is known of its habitat values, but where it is abundantly established bidgee-widgee can become a nuisance.

Control and management: the shallow-rooted plants can be readily pulled up by hand, preferably well before seedheads mature. Bidgee-widgee is occasionally included on planting lists of suitable species for created wetlands, but is an abundant and widespread plant in south-eastern Australia that needs no human assistance to spread. It should not be deliberately planted except in areas where both foot traffic and livestock are permanently excluded. People working in any way among seeding plants of this species should take care to remove all of the firmly attached seeds from clothing, and dispose of these where they cannot germinate.

Azolla (Azollaceae)

Fairy ferns are small, floating plants that can cover extensive areas of wetlands and farm dams in high nutrient situations; in sunlit places their red to brick-pink colouration can be recognised from a distance (see plate 15).

Origins: both species are indigenous and widespread, as well as on other continents, though ferny azolla is mainly found in warmer areas of the mainland.

Uses: all *Azolla* species carry a nitrogen-fixing symbiotic cyanobacterium (*Anabaena azollae*), and also accumulate other nutrients in amounts far beyond their immediate needs, so they have been extensively researched overseas as inexpensive, natural fertilisers.

Preferred growth conditions: Pacific azolla (*A. filiculoides*) grows primarily in cooler weather, while ferny azolla (*A. pinnata*) thrives best in warmer conditions, and only grows vigorously in summer in the southern parts of its range.

Confusing species: none. The rootlets of ferny azolla are finely divided compared to the lank strands of Pacific azolla, and it grows in a flat, pyramidal pattern lying flat on the water, while Pacific azolla is more eccentrically branched and may form heaped mounds on the surface in nutrient-rich situations. Both species turn red in full sun or when heat-stressed but can be separated by their colour alone, Pacific azolla being a brick-pink colour while ferny azolla is closer to scarlet.

Environmental effects: fairy ferns are a common component of the flora of many wetlands, and when present in moderate quantities are probably important in natural recycling of nutrients, as well as providing food for ducks, some other waterbirds, and almost certainly a wide range of smaller animals including diverse invertebrates. In cooler summers when Pacific azolla continues to grow prolifically through the season, it can create seriously anoxic conditions under the surface, and if it dies off *en masse* conditions may become anaerobic, resulting in the death of any aquatic animal that cannot breathe air directly.

Control and management: I chose the physical structure of Pacific azolla for a research project in second-year botany, and was impressed by the thick, waxy covering of the leaves, which no herbicide or wetting agent can ever breach effectively. Although some herbicides are supposedly effective against fairy ferns, it is my belief that the very few successes recorded are the result of spraying when the plants were already under stress from other causes, and were about to die away in any case. This has long been a problem in cultivation of 'seed' stocks for use as fertilisers in Asia, maintained as starter cultures in closed greenhouses to minimise invasion by a wide range of pests from viruses to herbivorous insect larvae such as rat-tailed maggots. Experimenting with rudimentary biological control along such lines, I have had some success introducing plants from obviously unhealthy or dying populations to thriving populations, with well-defined and spreading patches of decay appearing among the healthy plants from handfuls of (presumably) diseased plants thrown in from the shoreline. Although *Azolla* is still present in all such locations I have been able to revisit, it has never returned as a blanketing weed again to my knowledge.

Ceratophyllum (Ceratophyllaceae)

Hornwort or foxtail (*C. demersum*) is a floating, submerged plant with whorls of narrow leaves forming long, branching chains in waters up to several metres deep. Unlike most flowering plants with this growth habit, hornwort usually remains free-floating, though it may sometimes loosely attach itself to a silt substrate.

Origins: one of the most naturally widespread species of aquatic plants, found virtually worldwide, but as it produces little viable seed it is not clear how it spreads (see plate 13).

Uses: an ornamental aquarium or pond plant.

Preferred growth conditions: clear, ideally slightly alkaline waters maintaining a reasonably constant temperature, from deep, cool springs to slow-moving tropical streams.

Confusing species: resembles the submerged growth of some *Myriophyllum* species, but unlike these never grows above the water's surface; also superficially similar to some *Chara* species.

Environmental impacts and values: hornwort is a valuable habitat plant used as shelter by many types of tadpole, and a spawning medium by some surface-dwelling fishes. However, when present in large quantities it may taint the water with a fishy smell, and has been known to choke irrigation channels and the turbines of hydroelectric plants. It is a declared noxious weed in the Northern Territory, where perennially warm conditions are ideal for its growth through much of the year, and is a prohibited weed in Tasmania.

Control and management: hornwort is relatively sensitive to abrupt changes in water conditions, which has restricted its usefulness as an ornamental plant. Plants may disintegrate if water conditions change abruptly, for example if temperatures drop 10°C or more in just a few days, or if water conditions change from its preferred alkaline range to somewhat acid. Dredging out the bulk of the plant mass so only deeper-growing clumps remain, combined with increased turbidity from the dredging may also cause significant setbacks, as can drawdown during hot weather which can increase water temperatures rapidly.

Chara and *Nitella* (Characeae)

Despite their apparently complex organisation all stoneworts (*Chara* and *Nitella* species) are algae, so they never flower and have no roots, although they can attach themselves loosely to soft silt. They vary from coarsely divided, coralline species to densely matted types, and most of them require specialist knowledge for accurate identification (see plate 31).

Origins: there are many native species (most of them cosmopolitan) with fairly specific habitat and water quality requirements, but the more common species may be found in a wider range of habitats from farm dams to slow-moving, shaded creeks.

Uses: limited value as spawning medium for some fishes. Many species of *Chara* accumulate calcium (hence the name stonewort) and these may clear water in cloudy farm dams, where this has been caused by dispersive clays whose particles are so small they will not settle unless bonded into larger particles by lime.

Preferred growth conditions: the more common species prefer reasonably clear water from around neutral to somewhat alkaline, and are not fussy about temperatures – as would be expected for plants found naturally from Tasmania to the tropics.

Confusing species: superficially similar to *Ceratophyllum* and some smaller species of *Myriophyllum*, but no flowers are formed. Instead, stoneworts produce globular 'gametangia' which are their reproductive structures. In many forms of *C. corallina* these are such a bright orange in colour that they can be seen from a distance, in others they may be a close match to the stem colour.

Environmental impacts and values: the most widespread and adaptable species include coralline stonewort *Chara corallina* (sometimes also known as *C. australis*), feathery stonewort *C. fibrosa* and *C. vulgaris*, all of which can grow densely enough to taint water in farm dams: overseas they have also been known as muskgrasses for this reason. Sometimes stoneworts in farm dams may die off abruptly for unknown reasons, causing even worse water quality problems.

Control and management: sudden increases in water temperature or drops in pH, as have already been suggested for *Ceratophyllum*, can trigger mass mortality, but should be preceded by removal of as much of the stonewort as possible, to minimise water quality problems.

Cyperus (Cyperaceae)

Flatsedges include a number of grass-like plants which grow at the fringes of seasonally flowing irrigation channels, with variable flatsedge or Dirty Dora (*C. difformis*) regarded as a significant weed in rice fields. The much larger tall flatsedge (*C. exaltatus*) has also sometimes been regarded as a weed around supply channels, but this perennial plant needs permanently moist conditions to thrive, and has disappeared or become markedly less abundant in many parts of its southern range over the past decade of regular droughts.

Cyperus difformis.

Origins: variable flatsedge is a pantropical weed, particularly abundant in rice fields where it grows faster than the rice itself.

Preferred growth conditions: an annual with similar needs to those of rice (*Oryza sativa*), except that it can grow to maturity and set large quantities of seed months before the rice plants themselves mature.

Confusing species: some other smaller sedges may look similar before flowering, but the dark, compact flowerheads of variable flatsedge are distinctive.

Environmental effects: a weed of rice crops only, possibly with some value as a habitat plant, and potentially a useful, fast-establishing species that could be used to prevent establishment of less desirable introduced species in disturbed situations and created wetlands.

Control and management: the triangular stems make it easy to tell this plant from rice from an early stage, and control should be focused on physical removal of seedlings before they have a chance to set seed as the rice plants will outcompete Dirty Dora once they are large enough. Control of the all-too-abundant flatsedge plants usually found along the irrigation channels that feed the rice fields would also help to reduce the number of seedlings present in later crops.

Damasonium (Alismataceae)

Starfruit (*D. minus*) is an endemic relative of water plantains (*Alisma*), discussed elsewhere, and looks very similar though it is considerably smaller with distinctive fruits a little like a starry cogwheel at first glance.

Origins: found across much of Australia in suitable habitats, but not beyond.

Preferred growth conditions: as for *Cyperus difformis*.

Confusing species: can be confused with young water plantain plants and the native swamp lily (*Ottelia ovalifolia*; see also *Aponogeton distachyos* for more information) until the flowerheads appear.

Environmental effects: a minor weed of rice fields.

Control and management: see comments under *C. difformis*; however, in southern Australia starfruit is more in need of conservation than eradication, as many of its habitats have been dry for a decade or more, so rice fields may actually be a bulwark against its localised extinction in some places.

Duckweeds (Lemnaceae)

Duckweeds are tiny, thin-leaved floating plants that are present in most wetlands in small numbers, and may form a blanketing carpet over the surface in high nutrient situations, but are unlikely to grow so densely as to cause markedly anoxic conditions (see plate 16).

Origins: most duckweeds in Australia are indigenous, although there are a few localised records of European duckweed (*Lemna minor*), presumably introduced by the water garden trade.

Uses: these highly specialised and greatly simplified plants include many fast-growing, protein-rich species with considerable potential as food plants not only for livestock, but in some cases as a protein-rich food for humans as well. Egg-of-the-water (*Wolffia arrhiza*) is so closely related to the native *W. australiana* that the two species were long considered to be the same, and the Asian species has been recorded to produce around 10 tonnes per hectare (dry weight!) over its nine-month growing season; see Romanowski (2007) for a more detailed discussion of duckweed properties and growth patterns.

Preferred growth conditions: warm, nutrient-rich surface layers of smaller water bodies.

Confusing species: mainly other duckweeds, but for most purposes accurate identification is not necessary. Ignoring ivy-leaf duckweed (*L. trisulca*) which has never been recorded as being weedy, the native genera are *Spirodela* and *Landoltia* with several thread-like leaves hanging from each leaf, *Lemna* with a single root, and *Wolffia* species that are so small they don't need roots at all, but look like minute green globes that have been sliced in half.

Environmental effects: duckweeds can become a minor problem in nutrient-rich waters, but are otherwise a widespread and significant component of many natural and undisturbed wetlands.

Control and management: identify and reduce the source of the nutrients causing the problem.

Eleocharis (Cyperaceae)

Tall spikerush (*E. sphacelata*) is a robust sedge that can take over large areas of wetlands and smaller dams (see plate 8).

Origins: a widespread indigenous species also found in New Guinea.

Uses: significant nesting habitat for some waterbirds, and also recorded as an edible species by one dubious source, probably as a result of confusion with the native form of Chinese water chestnut (*E. dulcis*).

Preferred growth conditions: nutrient-rich waters up to 3 metres deep, though I have seen it growing vigorously from the bottom of a 6-metre deep dam that was regularly drawn down for irrigation purposes.

Confusing species: several other larger *Eleocharis* species, though these can usually be separated by their more restricted range and preference for warmer, shallower waters.

Environmental effects: a colonising species that can form extensive monospecific stands especially in farm dams; some exotic species in this genus are known to be allelopathic which presumably explains why it is unusual to find other plants growing with this species.

Control and management: control is not generally needed in most natural situations, but it should not be deliberately planted in created wetlands where a diverse range of habitat types are desired.

Epilobium (Onagraceae)

Willowherbs are mostly perennial herbs that spread by prolifically produced, wind-blown seed, colonising a wide range of disturbed habitats from recently burnt areas to drying lake beds according to species. Two exotic wetland species are discussed in the next section, but two indigenous species are included here as they can grow very vigorously under some conditions (see plate 29).

Origins: both robust willowherb (*E. billardierianum*) and hairy willowherb (*E. hirtigerum*) are indigenous, though they are also widespread beyond Australia.

Preferred growth conditions: damp to wet situations including drains and other disturbed habitats. Over the past decade or so of drought in southern Australia these two species have become among the most abundant colonising plants of drying lake floors.

Confusing species: many other *Epilobium* species; if there is any doubt whether an unfamiliar species is indigenous or introduced, contact your state herbarium or weed authorities for accurate identification.

Environmental effects: none recorded; sometimes a food source and refuge for a moth which also feeds on grapevines.

Control and management: willowherbs are largely a perceived problem rather than a real one, partly because plants producing seed can look remarkably unattractive, and partly because they spread rapidly into disturbed or recently exposed habitats, including created wetlands.

Hydrilla (Hydrocharitaceae)

The indigenous though sometimes weedy *Hydrilla verticillata* is briefly discussed with its introduced relatives *Elodea* and *Egeria*, but is a valuable submerged habitat plant across northern Australia and is not considered as a weed in this book (see plate 23).

Lachnagrostis (Poaceae)

Blown grasses, sometimes referred to as fairy grass, include several native species that have been increasingly treated as a weed problem as they spread over lake floors exposed by drought in southern Australia. The most widespread and problematic species is common blown grass (*L. filiformis*; see plate 4).

Origins: a native plant found through most of Australia, and also possibly beyond to New Zealand and Polynesia.

Preferred growth conditions: damp to wet, disturbed soils which may be quite saline.

Confusing species: other *Lachnagrostis* species, particularly *L. aemula*, though these are far less likely to be weedy.

Environmental effects: probably a significant habitat plant offering both shelter and seed to a range of wetland and some terrestrial birds, from ducks and finches to cisticolas. The loose, open, windblown flowerheads may build up in large volumes against buildings and other structures, and are sometimes regarded as a fire hazard in these conditions.

Control and management: despite the visually impressive volume of a mass of blown grass flowerheads, there is actually very little fuel present, and the heaped heads can easily be compacted down to a small volume with a roller, or by dampening and compressing.

Landoltia (Lemnaceae)

See Duckweeds.

Lemna (Lemnaceae)

See Duckweeds.

Leptochloa (Poaceae)

The various beetle- and cane-grasses include two fairly widespread but minor weeds of rice crops and irrigation canals.

Origins: the wetland species of this genus are indigenous, and are widespread across the warmer and drier parts of the mainland. See also *Diplachne* in the next section.

Uses: probably significant seed sources for a range of finches and related terrestrial birds.

Preferred growth conditions: brown beetle-grass (*L. fusca*) is a short-lived, salt tolerant species usually found in seasonally flooded areas which may be quite deep. Umbrella canegrass (*L. digitata*) is a longer-lived, slow-creeping grass with fairly upright canes to 2 metres that may be abundant along irrigation channels, but is better adapted to seasonal drought and drier conditions than many less desirable plants, that also are more likely to reduce flows still further by transpiration.

Confusing species: other *Leptochloa* species, though the two specifically discussed are the ones most likely to be perceived as weedy.

Environmental effects: only regarded as weeds along irrigation canals and seasonally flooded situations such as rice fields. It has been suggested that umbrella canegrass provides shelter to feral pigs, but it is unlikely to be large or dense enough to make much difference, and it is more likely to be the channels this grass grows along that are a significant factor in the spread of pigs through drier areas.

Control and management: the comments under *Cyperus difformis* earlier in this section apply equally to *L. fusca*, while *L. digitata* would probably best be regarded as a more desirable coloniser of irrigation channel banks than other more invasive and water-hungry plants.

Marsilea (Marsileaceae)

Rainbow nardoo (*M. mutica*) is a four-leaved fern that superficially resembles some clovers, and spreads rapidly over nutrient-rich soil in waters up to around half a metre deep. Although it is indigenous along the east coast of Australia, as far south as eastern Victoria, it has also been introduced into a coastal stream in southern Tasmania about 50 km south of Hobart and is now a prohibited plant in that state (see plate 12).

Uses: all native nardoos can be significant habitat plants in ephemeral wetlands within their natural range. Their sporocarps (formed most abundantly when water levels fall) have been cooked by Aboriginal peoples, but are reputed to need special preparation.

Preferred growth conditions: tolerates a wide range of conditions from the tropics to cold temperate climates, where it becomes dormant in winter, and is known to survive overseas even if the water's surface freezes. It will also survive prolonged drought, as long as the root mat is covered in a layer of undisturbed soil.

Confusing species: several other native species of nardoo; however, *M. mutica* often has much more pronounced reddish or ochre markings on the leaves, and plants growing in shallow water or on wet soils do not develop the furry coating of most other nardoos.

Environmental effects: on the mainland this can be a significant habitat plant, and the densely packed stems supported by the floating leaves provide shelter for smaller fishes and tadpoles. Regarded as a nuisance weed in Tasmania, it is unlikely to spread beyond the stream where it is already established, unless deliberately planted.

Control and management: a habitat plant and not a weed on the mainland, though the Gippsland (Victoria) population is regarded as possibly introduced by some people. This seems unlikely as it was recorded as early as 1929, long before there was much interest in growing or gardening with native wetland plants, and the species is not uncommon from the New South Wales border northwards, extending inland as well as along the coast. It is significant that the southernmost mainland population has not spread appreciably since it was first recorded.

Melaleuca (Myrtaceae)

Paperbarks are perhaps the most characteristic, widespread and abundant indigenous wetland shrubs, with diverse species found around all of Australia's coastal regions, with more terrestrial species extending well inland (see plate 12).

Uses: a significant type of wetland habitat from Tasmania to the tropics, only included here as potential weeds because of concerns that garden escapees are spreading some species beyond their natural range.

Preferred growth conditions: the true wetland species grow in areas which flood for long periods of time, drying out for part of the year, with their seed germinating on drying mud. There are distinctive suites of paperbarks adapted to specific climates, for example several in tropical Australia extending as far as India, with mostly smaller-leaved species in southern Australia.

Environmental effects: some wetland paperbarks are well known to be weedy when introduced outside their native range including broad-leaved paperbark

(*M. quinquenervia*), one of the most invasive introduced plants in Florida, USA. However, within its natural range from around Sydney northwards to New Guinea and Indonesia, this is such an abundant and ecologically important wetland plant that it is hard to imagine there is any suitable habitat it has not yet colonised. Further south, *M. parvistaminea* and the salt-tolerant *M. halmaturorum* have already naturalised beyond their natural ranges, almost certainly from garden escapes.

Control and management: paperbarks are only likely to become weeds if planted outside their natural range. Even if it is not known how serious this problem is likely to become, as a precautionary measure restrictions should be imposed on sale or distribution of particular species in areas where they don't belong. Although flooding is an important part of their seasonal cycle, even the true wetland species will weaken and gradually die if kept permanently flooded, and this would be the simplest control measure where it can be applied.

Myriophyllum (Haloragaceae)

There are many indigenous species of water milfoils, most of them growing in relatively shallow waters, with finely divided underwater foliage that becomes stiffer and more compact above the surface. Two closely related species are occasionally perceived as weedy: red water milfoil (*M. verrucosum*) and lake water milfoil (*M. salsugineum*).

Emergent growth of *Myriophyllum verrucosum*.

Origins: nearly all water milfoils in Australia are indigenous, but see also the discussion of *M. aquaticum* in the next section. Red water milfoil is found across all of mainland Australia in a wide range of conditions, while lake water milfoil is a more southern species often found in coastal areas.

Preferred growth conditions: the two species discussed grow most vigorously in clear waters, and lake water milfoil can be found up to 3 metres deep in ideal situations such as the calcium-rich waters of springs in south-eastern Australia.

Confusing species: some other water milfoils, though few of these are likely to be weedy except in high nutrient situations. Red water milfoil has relatively soft submerged foliage which smells rather fishy and collapses when lifted out of water, and the small, pink to red emergent leaves are deeply toothed and divided. The submerged foliage of lake water milfoil is firmer so the stems don't alter much in appearance when removed from the water, and the short, emergent stems with slightly notched leaves are only produced when the plant is flowering.

Environmental effects: many water milfoils are significant habitat plants, providing shelter from predatory birds for tadpoles, smaller fishes and diverse invertebrates. It is not clear whether these two species help to clear water, or just grow best in waters that are clear to begin with.

Control and management: red water milfoil grows rapidly in warm conditions and can fill farm dams or even small lakes in just one or two years, but is usually a short-lived problem and may disappear as abruptly it first appears. Lake water milfoil can be a more perennial problem, sometimes growing thickly enough to interfere with boating and fishing. Drawdown should provide an effective short-term cure for a few years for either species.

Najas (Najadaceae)

Prickly water nymph (*N. marina*) is a sharp-edged submerged plant which sometimes grows so densely in open waters that it interferes with fishing, and as it is extremely unpleasant to the touch is greatly disliked by water skiers.

Origins: occurs naturally across warmer regions of most continents, including Australia.

Preferred growth conditions: fresh to slightly saline, clear waters up to 3 metres deep.

Confusing species: none.

Environmental effects: possibly a useful habitat plant, but usually uncommon so that little has been recorded in this area.

Control and management: the deep, open water habitat makes this species almost impossible to manage. However, it has been suggested that increasing numbers of carp have killed it off in at least one Australian lake, so a combination of cutting, removal and increasing turbidity to prevent regrowth may work on smaller infestations.

Nitella (Characeae)

See *Chara*.

Phragmites (Poaceae)

Common reed (*P. australis*; see plate 7) is one of the most familiar native aquatics, a tall, bamboo-like grass reaching around 3 metres tall which is a dominant plant of coastal wetlands and slow-moving sections of river. The closely related, also indigenous *P. vallatoria* is generally even larger, but is associated with estuarine or even tidal areas in the tropics so it is not discussed here.

Origins: common reed is a cosmopolitan plant occurring naturally from some of the coldest parts of the northern hemisphere to Tasmania.

Uses: used worldwide for diverse purposes from thatching to paper-making, and commercially cropped for this purpose in parts of Europe, with edible young shoots and starchy rhizomes.

Preferred growth conditions: permanent waters are preferred, but common reed also grows in slightly saline tidal zones and places that dry out seasonally, though it is usually much smaller where water is not available throughout the year.

Confusing species: the introduced giant reed (*Arundo donax*), discussed in the next section, but common reed is easily recognised by its much more open and feathery flowerheads.

Environmental effects: natural stands of common reed provide a limited type of habitat, which may be fairly species-poor in terms of the variety of animal life present, but reeds are useful breeding and feeding grounds for some types of invertebrates as well as bitterns and reed warblers.

Control and management: only regarded as a weed on poorly drained pasture lands, where reeds may be significantly weakened by slashing during the drier times of year. If this is done not long before their dormant period over the cooler months they will be weakened still further, and cattle will readily eat the soft, spring shoots to complete control measures.

Spirodela (Lemnaceae)

See Duckweeds.

Utricularia (Lentibulariaceae)

Floating bladderwort (*U. gibba*) is a slender carnivorous plant that can form dense mats in small, shallow ponds or make up part of the fringing vegetation in larger water bodies. It is often only noticed when its distinctive, bright yellow flowers appear (see plate 12).

Origins: a widespread, variable species probably native across the warmer parts of many continents, including much of mainland Australia.

Preferred growth conditions: warm, shallow waters with abundant microscopic life, from which it derives many of its nutrient requirements.

Confusing species: some other smaller aquatic bladderworts, though no other indigenous species is as likely to become weedy.

Environmental effects: a common weed in aquariums and botanical gardens worldwide, sometimes a problem in smaller water bodies where it can form a choking blanket that reduces light and oxygen penetration. Introduced populations have been recorded as far south as Melbourne, probably as a result of water garden or aquarium introductions.

Control and management: left to its own devices, this species may have spread into the most southerly parts of Australia without assistance, but the current urban records are almost certainly inadvertent transplants. It is likely that floating bladderwort will continue to be spread over relatively small distances by waterbirds, as even the tiniest, thread-like stem will grow, and is not likely to be noticed until it has already established. In smaller pools, a combination of removal of most of the floating mat, combined with a well-mixed dose of copper sulfate at 1 gram per 200 litres of pond water is reportedly effective in control. However, the copper sulfate rapidly bonds into silt so it may become ineffective within days.

4

Compendium of weeds

Introduction to the encyclopedia

Writing a balanced account of wetland and aquatic weeds in Australia is no easy task, and in the encyclopedic entries that follow many species have been treated unequally. On the one hand, the information available on some of the worst weeds is so abundant that only an outline of the most important aspects of their biology and impacts within Australia could be included, without doubling or tripling the size of this book. These well-documented species are summarised along with the most useful sources, references, manuals and contacts for their control, and for most such weeds there are national programs and protocols that must be followed if we are to have any hope of containing them.

By contrast, many minor weeds and some plants that have yet to naturalise are covered in apparently disproportionate detail, because these may tell us more about what we need to be doing to prevent further outbreaks of new weeds, than retrospection over what has gone wrong in the past. These so-far minor problems include water garden and aquarium plants which are being widely distributed through Australia yet are already proven weeds elsewhere, new species which have arrived in this country under dubious circumstances, and plants that are likely to become increasingly weedy as global warming bites.

At another extreme, the claimed and actual impacts of some weeds are assessed from a slightly different perspective, especially in the context of their actual impacts on relatively undisturbed wetlands. Demonising plants that only have limited impacts on natural ecosystems is counterproductive, because this approach trivialises the importance of taking action against real and truly frightening weeds. While the author does not encourage taking a passive approach to weed control in our wetlands and waterways, there are many species we are likely to have to live with, and we need to be reserving our best efforts for those weeds that are shifting the balance of the natural world, or may do so in the future.

Grasses and related plants

Aquatic and wetland grasses of the family Poaceae are among the most widespread, abundant and, in some cases, serious wetland weeds. Dealing with them is made all the more complex because many of them aren't easy to identify when not in flower. They are considered here along with the closely related sedges and several other obviously related families, for more convenient comparison.

Grasses: the family Poaceae

Arundo (Poaceae)

Giant reed (*A. donax*; see plate 7) resembles a larger and more robust version of the native common reed (*Phragmites*, discussed in the previous chapter), and with its thick stems to around 4 metres in height is often mistaken for a bamboo. However, most true bamboos hold their leaves on branches coming out from the nodes (joints) along the stems, while the single, strap-like leaves of reeds are directly attached at the node. The relatively narrow and upright flower plumes of giant reed are very different from the more open and feathery flowerheads of common reed.

Origins: a European to western Asian grass that is widely cultivated, often in a variegated leaf form that is less vigorous than the typical green species form.

Uses: probably introduced as an ornamental.

Preferred growth conditions: grows in a wide range of soils, but prefers moist to wet conditions and places that may be flooded at times. Unlike bamboos, freshly cut stems used as garden stakes take root easily if kept moist in warmer weather, and it is likely that many feral populations have established from seed.

Confusing species: the more slender common reed (*Phragmites*), though this has a much more open flowering head than the narrower plume of *Arundo*.

Environmental effects: an invasive plant in many wetland situations in warmer areas, such as around Newcastle and some lakes in south-western Western Australia, and also more sporadically naturalised elsewhere. Plants growing in cooler temperate areas don't seem to be particularly invasive at the present time, but this could easily change with just a few degrees of global warming.

Control and management: there is nothing particularly attractive about the green form of giant reed, and it is unlikely that anyone but the most obsessive collectors of grasses would object to its being declared a weed. The smaller and less vigorous variegated form is more manageable, does not apparently set viable seed in most situations, and could be exempted from any restrictions but should still not be planted near wetlands. Bringing existing feral populations under control is a more difficult proposition, but as these usually grow on the drier fringes of wetlands they can usually be slashed late in summer or in autumn when the ground they grow on has dried out and become firm enough for heavy machinery. Even full-sized plants can be diminished dramatically by two or three slashings, so that they no longer flower and are more accessible for further control.

Cortaderia (Poaceae)

Pampas grasses (see plate 5) grow well on permanently moist to wet soils, forming large, wiry-leaved tussocks with prominent plumes of flowers held well above the foliage. The common species of pampas grass (*C. selloana*) with silvery, creamy or yellowish flowerheads is usually the largest at around 4 metres high when in flower, though this may be because there are more well-established plants of this species in cultivation. Pink pampas grass (*C. jubata*) with pink-tinted or purplish flowerheads is a more recent introduction that shows signs of spreading at an alarming rate. Both species can spread considerable distances, growing from their prolifically produced, wind-blown seed. A third species, the more recently introduced *C. richardsonii*, has already naturalised in Tasmania, in a climate to which it is biologically pre-adapted.

Origins: both of the more familiar and widespread species are native to South America, while *C. richardsonii* is from New Zealand.

Uses: grown purely as ornamentals, though not so widely planted these days because of their considerable size and increasingly obvious weediness.

Preferred growth conditions: most moist to wet soils, tolerating brief periods of flooding.

Confusing species: none.

Environmental effects: pampas grasses have been used as garden plants for so long that most people aren't consciously aware of them as an invasive blight, and in many cases have a liking for them, yet there is every sign that they could become one of the most invasive plants of temperate and subtropical wetlands. *C. selloana* is already a well-established wetland and environmental weed in warmer temperate areas, while *C. jubata* is more invasive in cooler areas including Tasmania and does not need cross-pollination to produce seed. During periods of drought all pampas grasses can be highly flammable, and old tussocks contain a considerable amount of dead, dry fuel. Pampas grasses have no habitat values in Australian conditions, and are already declared noxious in some parts of Australia.

Control and management: pampas grass species should be banned and eradicated wherever possible as a priority, before they become as widespread a problem as the willows (*Salix*). Even in their native environments they can cover extensive areas, and if allowed to spread further in southern Australia these gigantic grasses could potentially become the dominant vegetation of wetland fringes over a wide range of climates and soils, spreading still further as climate change and drying lake beds make an ever-increasing array of suitable habitats available to them.

In wetlands where water levels can be controlled seasonally, flooding for several weeks or more can be used as a control method. In all other situations, given that appropriate herbicides can be applied directly into the living crown of these invasive plants, this would be an acceptable control method with far less impact on the wider wetland environment than the plants themselves.

Given the large size of these grasses, and that the total retail sales of all species of pampas grasses through the nursery trade are likely to be well under $50 000 annually within Australia, the market value of these plants is miniscule in comparison to the environmental disaster they are likely to develop into. Nor are they being planted so readily by gardeners these days: one popular garden program a few years ago even featured a segment on destroying mature clumps for the many gardeners who are increasingly coming to regard them as a waste of space in the contemporary garden.

Diplachne (Poaceae)

American beetlegrass (*D. uninervia*) is similar in general appearance to several indigenous grasses in the related genus *Leptochloa*, described earlier as minor weeds of rice fields. However, the spikelets are shorter than in any of the native species, as is the lemma, a papery bract at the base of each spikelet.

Origins: native to Northern America and recorded as naturalised in New South Wales near Homebush Bay, with an isolated inland record from rice-growing areas.

Preferred growth conditions: annual or short-lived coloniser of disturbed, shallowly flooded soils.

Confusing species: native beetlegrasses, particularly *D. parviflora*.

Environmental effects: a major rice field weed in California, but it is not known whether it is likely to be invasive in relatively undisturbed wetlands.

Control and management: every effort should be made to eradicate the localised Homebush Bay population as a pre-emptive move (assuming this has not yet been done at the naval base where it has been recorded); the inland record also needs to be followed up to check whether this species is still present elsewhere.

Echinochloa (Poaceae)

Barnyard grasses include a few indigenous species as well as some widespread and weedy introduced forms. In southern Australia the best-known species is barnyard grass (*E. crus-galli*; see plate 4), a prolific weed of drains and disturbed situations such as rice fields. It is regarded as one of the worst agricultural weeds in the world, although it is not usually found in undisturbed wetlands. Several other species that are grown as millets are also locally naturalised in places, but whether these will persist in the long term is not known; the taxonomy, common names and relationships of these are somewhat confusing and need not be considered here. Aleman grass (*E. polystachya*) is a perennial tropical species reaching up to 3 metres in height, and is relatively easily recognised as it spreads by runners – unlike the annual species.

Origins: barnyard grass was probably introduced from Europe, although it is so widespread that its true origins are uncertain, and is now found across most of Australia. Aleman grass was introduced as a pasture grass from the Americas and was still being promoted as a suitable forage grass for seasonally inundated soils until well after its weed potential had been made clear, and it is now thoroughly naturalised in tropical Queensland and the Northern Territory.

Uses: many of the minor weedy species of this genus are grown as millets on wet ground (Romanowski 2007), and even the seed of common barnyard grass is edible though slightly bitter.

Preferred growth conditions: seasonally flooded soils and damp areas, growing most vigorously in nutrient-rich situations.

Confusing species: the indigenous awnless barnyard grass (*E. colona*) is also regarded as a weed of moist, disturbed ground but is rarely aquatic, and does not have the short bristles (awns) found on barnyard grass. Most other native *Echinochloa* species are tropical to subtropical with conspicuous bristles, which give their seedheads a heavily bearded appearance, and are annuals forming distinct clumps, unlikely to be confused with the taller Aleman grass which spreads by rhizomes (runners), and has been reported to grow in waters up to 2 metres deep.

Environmental effects: despite its wide distribution and vigorous growth, barnyard grass is rarely found in natural wetlands. Aleman grass is recorded from around tropical billabongs and even found established in floating mats of vegetation, and its ability to invade relatively undisturbed wetlands and form dense stands under such competitive conditions suggests that it has the potential to become a serious environmental weed.

Control and management: as barnyard grass is an annual, removal or destruction of plants before they set seed will often control it, particularly in places where it has only just appeared. Seed will germinate for another year or two afterwards, but the seedlings don't compete well with indigenous plants. Aleman grass does not apparently produce seed in Australian populations, but spreads from broken off sections of rhizome – removal by hand in the early stages of an infestation may be a possible control option. Spraying is unlikely to be successful except where large areas have already been covered by this grass, and is likely to also affect indigenous plants which may be helping to restrict the spread of the weed.

Glyceria (Poaceae)

Several exotic sweetgrasses are naturalised in southern Australia, and while manna grass (*G. declinata*) and the similar-looking *G. plicata* don't seem to be regarded as serious problems at present, reed sweetgrass (*G. maxima*; see plate 5) is an aggressive weed that can block and even redirect the path of slow-moving streams, changing the nature of entire habitats.

Origins: originally from Europe and northern Asia, now found patchily through cooler parts of eastern Australia.

Uses: the seed of many sweetgrass species is edible, though it is unlikely that they were introduced for this reason. Reed sweetgrass was originally introduced as a pasture grass for wet soils but is seasonally dangerous to stock as it produces cyanide compounds in spring.

Preferred growth conditions: waterlogged soils and shallow, relatively permanent waters including slower-flowing streams, growing most vigorously in nutrient-rich situations.

Confusing species: the two native and non-weedy species of sweetgrass are usually narrower-leaved than the weedy introductions, but unless you are familiar with their distinctive overall appearance identification should be based on the shape, length and other characters of the spikelets.

Environmental effects: reed sweetgrass forms dense, monospecific stands that alter water flow patterns. It is largely avoided by native animals, possibly because of its seasonally toxic nature. It is a fast-spreading weed in the more open, sunlit sections of the river flowing 1 kilometre from where I live, taking over more of the backwaters on the grazed areas along its course every year. I have experimentally created *Glyceria*-free pools of several square metres in some of the densest sections to see if these are colonised by fishes, frogs or aquatic invertebrates, but even the introduced plague minnow (*Gambusia holbrooki*) has not found its way to these, though it is abundant both up- and downstream of the worst sweetgrass infestations. This also suggests that sweetgrass is likely to be avoided by most indigenous aquatic animals.

Control and management: chemical sprays and mechanical removal have been suggested as treatments for this virulent grass, yet neither method has produced any worthwhile long-term results as the extensive rhizome system usually recovers quickly. Mechanical dredging is often difficult due to swampy surrounding ground, and even the smallest piece of rhizome that escapes downstream can potentially generate new infestations, often in out-of-the way places where they remain unnoticed until too late. Spraying is usually only partly successful, and even if an area is successfully cleared of this weed, the seed is likely to continue germinating for several years afterwards so follow-through of initial control measures is essential. The worst infestations are usually in open areas such as backwaters on cleared pastureland, and the local joke where I live is that it is best to leave nearby willows in place as they are the only plant reed sweetgrass won't grow under. In turn, this suggests that rehabilitation of areas infested by this grass would best be treated as a long-term project, planting appropriate indigenous species that overhang the water and ultimately shade the weed out.

Hainardia (Poaceae)

At the present time, common barbgrass (*H. cylindrica*) is a relatively minor, though sometimes locally abundant, weed in poorly drained, saline soils.

Origins: southern Europe, now also scattered across much of non-tropical Australia.

Preferred growth conditions: heavy, often saline soils including the fringes of coastal wetlands, growing mainly through autumn and winter.

Confusing species: stunted plants in harsh situations or which are regularly mowed may be confused with the native mat grass (*Hemarthria uncinata*), which also may grow in similar situations.

Environmental effects: at present most abundant in cleared, disturbed situations, though barbgrass will also colonise relatively undisturbed wetland fringes in any open spaces between established plants.

Control and management: no specific measures or treatments have been suggested for this species as it is not usually so abundant as to present an obvious threat.

Holcus (Poaceae)

Yorkshire fog grass (*H. lanatus*) is primarily a terrestrial grass but can also be a persistent weed in disturbed wet places, particularly in created wetlands.

Origins: Europe, possibly introduced as a pasture species and now widely naturalised in south-eastern Australia.

Preferred growth conditions: moist to wet, reasonably nutrient-rich soils; established plants will tolerate long periods of flooding as long as they have enough of their root mass established above soil.

Confusing species: none when in flower.

Environmental effects: negligible in natural wetlands, and may be grazed upon by herbivorous birds such as swans.

Control and management: fog grass does not compete effectively against true wetland plants and will usually die out over time, even in created wetlands.

Hymenachne (Poaceae)

Olive hymenachne (*H. amplexicaulis*; see plate 26) is a weedy introduced pasture grass that may reach well over 2 metres when growing in shallow water over rich soils. Deliberately introduced as recently as 1988 as a pasture grass, it has proven to be so weedy that some of the same agencies responsible for its original release are now struggling to try to keep it contained within manageable bounds, just two decades after it exploded into the wider environment.

Origins: introduced from South America, now widely naturalised in tropical Queensland where attempts are being made to contain it to a relatively narrow strip of the eastern coast between Cooktown and a little north of Brisbane. Isolated populations (which have already appeared beyond this range as far south as northern New South Wales and all the way up to the tip of Cape York) are targeted for eradication while another containment line in the Northern Territory is intended to keep it from spreading beyond the relatively small area around Darwin where it has already established.

Preferred growth conditions: grows in still and slow-flowing waters up to 1 metre deep. It is invasive in a wide range of habitats from wet pastures to sugarcane fields, as well as blocking slow-moving creeks and sometimes relatively undisturbed natural wetlands. Although a true aquatic, this species only thrives in soils that are only flooded for some of the year, rather than in deeper, permanent waters.

Confusing species: the indigenous *H. acutigluma* is similar but not as large and vigorous growing. The two species can be separated by the heart-shaped leaf base in olive hymenachne, which clasps the stem so the two pronounced lobes on either side of the stem almost overlap. The comparably weedy para grass (*Urochloa mutica*) is found in similar habitats but is much smaller, and the compact, branched flowerheads of this species are very different from the more or less upright, rarely branched pokers of olive hymenachne that may reach 200 mm in length.

Environmental effects: a densely growing, creeping grass that can smother smaller plants and blanket considerable sections of streams, comparable to reed sweetgrass (*Glyceria*) in southern Australia in its impacts. In the tropics the growing season is so long that hymenachne grows much faster and for longer periods of time. The best documented impacts of this species on relatively undisturbed natural wetlands are in the Northern Territory, where this species has become a significant weed in just a decade, and is spreading rapidly into world heritage areas such as Kakadu National Park. Seed production in this species can be prolific with germination rates up to 98 per cent. As it can be spread by waterbirds, including magpie geese in the Northern Territory, preventing seed formation (especially in new infestations) should always be a priority in the broad-spectrum control of this species.

Control and management: this weed cannot be managed by any single individual except in the very first stages of an infestation. It is essential that any new, unrecorded or suspected outbreak should be reported immediately to state or territory authorities so it can be dealt with within the framework of the national management plan already in place. This can be downloaded from the Weeds of National Significance website (see 'Further information including websites'), and includes a good overview of the various control measures being considered or implemented. The paper *Control Methods and Case Studies*: Hymenachne amplexicaulis (2006) on this site is a useful summary of what is and can be done about this virulent weed, although it is already probably far too late to control. The focus is now on preventing its spread beyond a defined containment line.

There is little sign that control of established outbreaks of this species will be possible with chemical controls, and there are significant restrictions on the use of the very few that are legally allowed to be used on this species, for reasons discussed elsewhere. Biological control of hymenachne is being considered. However, this has been complicated by the presence of its native relative (which was an endangered plant not all that long ago due to grazing by water buffalo), and there are also concerns that any insect or disease that attacks olive hymenachne could also damage sugar cane. To date, the most promising biocontrol agent is a bug (*Ischnodemus variegatus*) in Florida, which apparently only feeds on olive hymenachne.

Leersia (Poaceae)

Rice cutgrass (*L. oryzoides*) is a perennial grass of wet places, reaching 1.5 metres in height, and spreading by long, underground runners. The native swampgrass (*L. hexandra*) is also apparently extending its range and is now found as far south as Victoria, but is not presently regarded as a notable weed.

Origins: originally northern America, now patchily established in eastern Australia.

Preferred growth conditions: heavy, wet soils which may be shallowly flooded at times, though usually not for long.

Confusing species: the native swamp ricegrass (*L. hexandra*) is regarded as a valuable fodder plant, but is also sometimes a minor weed of irrigation channels; unlike rice cutgrass this species does not seem to produce mature seed often.

Environmental effects: to date, mainly a weed of irrigation channels in rice-growing areas of south-central New South Wales, with a few outlying records in Victoria and Queensland. There are some indications that this species has become less abundant during recent drought years as water quotas for rice growing have been increasingly restricted, but its potential as a weed in relatively undisturbed wetlands is not known.

Control and management: only regarded as a relatively minor potential weed at the present time, so little relevant information on control in Australian conditions is available.

Panicum (Poaceae)

Torpedo grass (*P. repens*) is a moderately tall grass to around 1 metre or more in height, tolerating flooding at times but usually found on higher and reasonably well-drained ground.

Origins: the Mediterranean area, now naturalised patchily from central New South Wales to southern Queensland.

Preferred growth conditions: disturbed, often sandy soils, potentially spreading by runners along drainage canals and into seasonally wet pastures.

Confusing species: various indigenous *Panicum* species; whatever reputation torpedo grass has as a potential wetland invader may be partly due to confusion with the native swamp panic (*P. paludosum*), a true aquatic species with floating stems, found from northern New South Wales northwards into Asia.

Environmental effects: a weed of pasture lands and disturbed soils rather than wetlands.

Control and management: as torpedo grass does not appear to produce much seed in Australia and grows primarily on firmer ground where machine access is usually possible, it could potentially be managed by strategically timed slashing. Deeper cultivation just breaks up the rhizomes, which will usually regrow.

Paspalum (Poaceae)

Water couch (*P. distichum*; see plate 6) is a mat-forming grass that may form sprawling mounds at the water's edge, smothering smaller species already growing in the shallows, and will also grow in deeper but more ephemeral waters.

Origins: regarded as native in many tropical and warm-temperate countries, and until fairly recently as a widespread indigenous plant other than in Tasmania and the Northern Territory. In recent years this species has increasingly been treated as a possible introduced weed.

Uses: if indigenous, water couch is a minor habitat plant providing shelter for some migratory snipe and aquatic invertebrates, though these will shelter as readily in many other low-growing plants of seasonally wet places.

Preferred growth conditions: permanently moist to wet soils, growing most vigorously in high-nutrient, disturbed situations such as created wetlands.

Confusing species: the similar-looking saltwater couch (*P. vaginatum*) is also regarded as a possible introduction, and is found along the coastal regions of all mainland states in brackish or saline soils which may be wet at times, but are not usually flooded. Many other more terrestrial *Paspalum* species, both native and introduced, are found in poorly drained situations and may be weeds in drains and along irrigation canals.

Environmental effects: regardless of whether this grass is native or introduced, it is not usually an important habitat plant though I have found extensive, possibly very old floating mats (anchored among common reed, *Phragmites*) being used by breeding waterbirds in some apparently pristine places.

Control and management: water couch is only likely to become seriously weedy in disturbed, relatively high-nutrient places, with the roots established in wet soils along the fringes supporting a floating mat which may extend many metres from shore in still waters. This particular growth habit is unlikely to be entirely natural, but mechanical control by cutting the floating mat away from the anchored section on shore with a backhoe or excavator should allow other plants to regrow in the shallows near the shoreline.

Phalaris (Poaceae)

Several species of canary-grass have been introduced as pasture plants, of which Toowoomba canary-grass (*P. aquatica*; see plate 15) is the most abundant and widespread. Although this is a common weed of drains and other poorly drained places, tolerating waterlogging for some of the year, it is primarily a weed of disturbed ground and rarely becomes a problem in natural wetlands. Many less common canary-grass species grow in similar situations and are usually regarded as weeds of intermittently flooded agricultural land, but reed canary-grass (*P. arundinacea*; see plate 6) may become increasingly weedy and has the potential to invade wetter areas if the recent behaviour of some southern populations is any guide.

Origins: reed canary-grass is believed to be native across much of the temperate northern hemisphere.

Uses: introduced as a pasture grass, though no longer widely planted.

Preferred growth conditions: moist to wet soils which may be flooded for long periods of time, in which case the plant grows more vigorously and is more likely to set seed.

Confusing species: other species of canary-grass, though the flowerheads of this species are distinctively branched.

Environmental effects: patchily distributed and not all that common at present. However, some populations in my local area have spread to cover several hectares over the past decade, possibly as a consequence of alternating years of drought and occasional flooding, suggesting that the species could become a more problematic weed under the increasingly variable climatic conditions predicted as a result of global warming. It may be that the most vigorous populations are encouraged by intermittent grazing, and there is no clear evidence that these would be likely to become invasive in less disturbed situations.

Control and management: ideally, any populations established near natural wetlands should be destroyed before they have the chance to spread; on agricultural land they are likely to be contained by grazing.

Polypogon (Poaceae)

Annual beardgrass (*P. monspeliensis*; see plate 4) is an annual weed of disturbed soils and saltmarshes, though it may also grow prolifically in freshwater situations for a year or two before other plants crowd it out.

Origins: Europe and northern Asia, now widespread and sometimes locally abundant across most of Australia.

Preferred growth conditions: wet but not waterlogged soils, tolerating periods of flooding. Although plants may be found growing in fresher reaches of tidal saltmarshes, their roots are usually on slightly raised tussocks of ground, where they may be flooded at times, but there is generally some oxygen reaching the roots.

Confusing species: none in comparable habitats.

Environmental effects: a minor wetland weed that may grow densely at times, but will usually become much less common within several years, except where continued grazing reduces competition from other plants.

Control and management: manual removal of all plants before they set seed is often possible, and following this up annually for the next two years should be enough to destroy the soil seed bank.

Spartina (Poaceae)

Cordgrasses (also known as rice grasses in Australia) were stupidly introduced to stabilise coastal mudflats in south-eastern Australia, in much the same era and spirit as the planting of marram grass (*Ammophila arenaria*) to stabilise sand dunes. The combined effects on natural coastlines have been dramatic, with marram grass creating steep, high-pitched dunes more prone to blowout during storms than those stabilised by native spinifex grasses (*Spinifex* spp), while cordgrasses retain silt and create tidally inundated stands which are of limited

Spartina anglica invading a relatively pristine mangrove swamp at left, and at right showing the angular, semi-upright habit including the distinctive columnar flowerhead.

value as habitat to most native animals. As a result, people living in areas where these two introduced grasses have become established may never have seen an unmodified Australian coastline. There are two related *Spartina* species, of which Townsend cordgrass (*S.* × *townsendii*) is a little smaller than rice grass (*S. anglica*).

Origins: Europe, now well established in parts of south-eastern Victoria and Tasmania, with new populations occasionally appearing spontaneously further afield.

Preferred growth conditions: nutrient-rich tidal mudflats, both species growing from fragments of the rhizome, or seed in the case of rice grass (which is much more likely to spontaneously appear elsewhere).

Confusing species: rice grass is a more vigorous and slightly larger mutant form of Townsend cordgrass, and has often replaced the latter where the two grow together. Neither is likely to be confused with native grasses in the habitats where *Spartina* grows, and they can usually be separated by size alone. In rice grass the leaves reach 45 cm long and 15 mm wide with larger flowerheads to 40 cm long, while in Townsend's cordgrass the leaves are only 30 cm long and 12 mm wide, with flowerheads to 25 cm.

Environmental effects: both species appear to be fast-growing competitors for indigenous saltmarsh plants, and also significantly alter or reduce biodiversity by changing mudflats into saltwater meadows, which few indigenous animals have any use for.

Control and management: all outbreaks of these two grasses are on public land, and any new outbreaks should be reported to the relevant state or local authorities. There are well-established protocols for control of established populations in both Victoria and Tasmania, as well as for control of new infestations.

Sporobolus (Poaceae)

Parramatta or rat-tail grass (*S. africanus*; see plate 6) in its various forms is a wiry grass tolerating a wide range of soils and conditions from fairly dry hillsides to seasonally wet soils.

Origins: different varieties from both Africa and Asia have naturalised in Australia, mainly in the south-east though there have been some reports from Western Australia.

Uses: introduced as a fodder grass but of relatively poor quality.

Preferred growth conditions: moist to wet soils including poorly drained agricultural land where it may spread rapidly from its prolifically produced seed, as livestock prefer to graze more palatable grasses.

Confusing species: there are many native species of *Sporobolus* so identification by a herbarium is recommended if there is any possibility of confusion, especially of stunted plants on relatively dry ground.

Environmental effects: primarily a weed of disturbed soils and agricultural land, with no habitat values for native animals – even swamphens won't eat it!

Control and management: usually a relatively minor weed, though it can eventually become dominant on disturbed sites; smaller plants in wet soils are relatively easily pulled out. As the places where this plant prefers to grow are usually only intermittently flooded, localised control of larger populations may be achieved by flooding for several months.

Urochloa (Poaceae)

Para grass (*U. mutica*) is a tropical weed of seasonally flooded places, reaching 2 metres in height though the sprawling stems may be twice this length. Apparently introduced a century ago in the Northern Territory, and long recognised as a serious environmental and agricultural weed, deliberate planting of this species has continued over so many decades that it is already well entrenched over perhaps half of its potential range in northern Australia.

Origins: tropical Africa (although some sources suggest it may be from Brazil), now spread over many regions of northern Australia and probably still expanding its range, with outlying populations found as far south as northern New South Wales.

Uses: introduced as a pasture grass through many areas of the tropics, invariably becoming weedy to some degree wherever it is grown.

Preferred growth conditions: seasonally flooded tropical wetlands (it is usually killed by frost), though also growing through floating mats of vegetation in more permanent waters.

Confusing species: several other grasses including some forms of *Paspalum* and also the native swamp rice grass (*Leersia hexandra*), though the flowerheads of para grass are comparatively small and sparsely branched for the overall height of the plant: around 20 cm long, with the longest branches up to 8 cm. The leaf sheaths are lightly covered by fairly stiff hairs with slightly swollen bases, unlike most other wetland grasses.

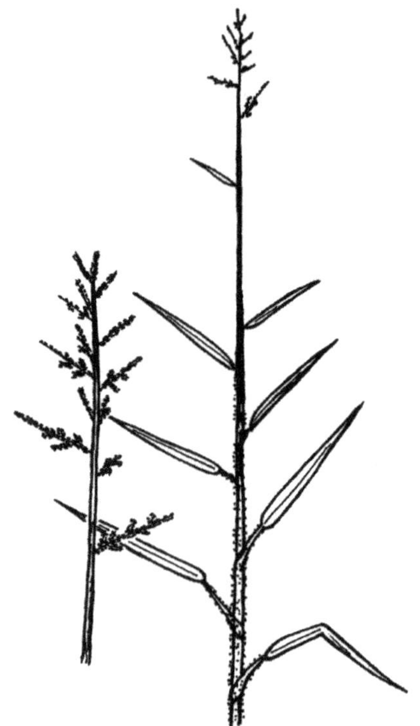

Urochloa mutica, showing the relatively small and sparsely branched flowerhead.

Environmental effects: an agricultural weed that has spread into diverse moist areas from sugar cane farms to drainage channels. It can also displace native vegetation in ephemeral wetlands, which appear to be an ideal habitat for this weed. Although not as invasive as olive hymenachne (*Hymenachne*) there is little doubt that this weed is changing the ecology of many wetlands where it has established.

Control and management: although this species appears to produce some viable seed in Australia (16 per cent viability is the highest estimate), it mainly seems to spread from runners and other fragments, so most populations are probably the result of deliberate planting. Once established it is almost impossible to control, and in seasonally dry wetlands in the Northern Territory it may dominate areas formerly covered by *Mimosa pigra*, recovering and spreading faster after fire than potentially competing indigenous species. The main problem with dealing with this species as a wetland weed is that its value as a pasture grass, combined with its long tenure as a weed, has conferred a certain degree of immunity from any detailed investigation of its impacts. Recent papers are more focused on whether this species significantly reduces biodiversity compared to indigenous grasses than on possible control methods, which at this late stage in its expansion would need to be coordinated at a national level. As these studies are primarily statistical evaluations that do not actually tell us much about what para grass does to specific animal and plant species, their superficially reassuring message may bear little relationship to what is happening in the wetlands themselves.

Sedges, rushes and other relatives of grasses

Carex (Cyperaceae)

Many species of these grass-like sedges are already established weeds in Australia, though few of them are true wetland plants. New Zealand species are particularly suspect as several of these have already naturalised in Tasmania, but several other *Carex* species have established localised populations in places and some of these may become significant weeds with time (see plate 9).

Origins: a wide range of ornamental *Carex* species has been imported from Europe, New Zealand and other parts of the group's global distribution prior to the 1990s, when a blanket ban on further imports from this genus was imposed. Several others have appeared since, possibly smuggled in the form of seed, but the fashion for grassy plants in gardens has waned and inspection of seed imports has become more stringent so it seems unlikely that smuggling will continue to be worthwhile.

Uses: introduced as ornamental plants only, with some possibly accidental imports.

Preferred growth conditions: there are hundreds of species of *Carex*, and even more cultivars. Many of these are relatively terrestrial in their requirements, but others will tolerate prolonged flooding, and some are true wetland plants. Of these, there are a number of species potentially invasive in a range of aquatic and water's edge habitats, some of which have already been introduced into Australia and are even being sold by nurseries. As these come from very different parts of the world, there is no general rule of thumb predicting where they are most likely to become weedy.

Confusing species: *Carex* species can be recognised easily when in flower, but at other times can be confused with a variety of true grasses (Poaceae).

Environmental effects: many *Carex* species are likely to become increasingly common weeds, as they continue to spread by seed in damp to wet places, and some grow so large and are so alien in appearance to indigenous species that they are capable of changing the very nature of southern Australian wetlands if they establish feral populations. Many of the New Zealand species are regarded as serious weeds even within their native range, which should give any nurseries propagating any of the species discussed below cause for thought.

Control and management: control of running species is difficult once they have established themselves on any scale, and prevention or immediate destruction of new infestations should be a priority. This could most successfully be done by a blanket ban on New Zealand *Carex* species throughout Australia; *C. buchananii*, *C. comans* in its various colour forms, *C. flagellifera* and *C. testacea* are already declared noxious in Tasmania, but are still widely available on the mainland. The giant 'niggerhead' sedge or makura (*C. secta*) of New Zealand wetlands is less well known but is being grown in at least one botanic garden, and as I have seen seed being illegally collected from these plants this species is likely to become commercially available in the near future.

Another very weedy *Carex* – which should be banned outright – continues to be widely sold as an indigenous plant under such names as *C. gaudichaudiana* 'Bruny Island form', *C. gunniana* 'Bruny Island form' and *C. glauca* (there is no such Australian species). This weed is actually the introduced European *C. flacca*. Already widely naturalised in Tasmania (yes, it is common on Bruny Island) and to a lesser extent Victoria, if nurseries continue to sell this invasive species mislabelled as a native it is likely to become still more widespread.

Chondropetalum (Restionaceae)

Thatching reed (*C. tectorum*) is an upright, virtually leafless, reed-like plant that is increasingly being offered as an ornamental plant through garden centres, even though it is neither attractive nor very different in overall appearance from some of its native relatives.

Origins: in its native South Africa this cordrush has long been used as long-lasting thatch, some of which is being imported into Australia.

Uses: a minor ornamental of limited appeal, which could easily be replaced by any of around a dozen similar-looking but indigenous native species from the same family.

Preferred growth conditions: seasonally wet to waterlogged places, but tolerating drought and nutrient deficient soils.

Confusing species: some other South African cordrushes and, to a lesser extent, various indigenous cordrushes.

Environmental effects: thatching reed has not yet naturalised in Australia but has the potential to do so. The germination rate of imported seeds I raised in the 1990s was around 90 per cent, and the resulting tray of plants was already producing seed two years later. Some of these were donated to the Sydney Botanic Gardens at their request, but the rest of the tray was buried in a 2.5-metre deep weed disposal pit. Transported by the truckload in South Africa for thatching projects well beyond its natural range, spilt seed of thatching reed has created naturalised stands along many roads, even in drier places than it would normally grow.

Control and management: most cordrushes are male or female, so fertile seed is only set if both sexes grow fairly close together. Plants available to gardeners should be propagated by division only, so that single-sex clones are all that is available.

Cyperus (Cyperaceae)

Drain sedge (*C. eragrostis*; see plate 8) is a useless though attractive weed that has spread so widely and rapidly, particularly in south-eastern Australia, that no real attempts to control it have been made for decades.

Origins: the Americas, but now thoroughly naturalised in many other places including much of temperate Australia.

Preferred growth conditions: one of the most adaptable wetland weeds, found in many types of disturbed and poorly drained habitats; stunted plants may even be found growing in the cracks of bitumen or concrete paths during wet summers. Often found in sites that are seasonally flooded and tolerant of waterlogged soils, but in more permanently wet situations drain sedge is usually replaced by more vigorous aquatics.

Confusing species: occasionally confused with native *Cyperus* species, particularly *C. exaltatus*, drain sedge has also been sold as the very different (native) floodplain plant *C. bifax*. The green flowerheads of drain sedge develop a rusty colour as they age, but obvious traces of green always remain – unlike in most comparable native sedges, the flowerheads of which are completely tan to bronze once mature.

Environmental effects: as far as is known this is a relatively innocuous weed, and some writers even suggest that it is ecologically comparable to some indigenous species. Drain sedge is primarily a weed of disturbed ground including created wetlands, and although plants may sometimes be found scattered through the fringes of relatively undisturbed wetlands, these are unlikely to spread far.

Control and management: easily controlled when it first appears by simply pulling up young plants, but within a year or two so much seed has usually been set that there is little prospect of permanent control. Pulling up of mature plants usually just exposes buried seed, which grows much more rapidly on the newly disturbed ground than most indigenous plants, so heavy mulching of such exposed situations for at least three years is recommended to prevent seed germination.

Cyperus – ornamentals and minor weeds

Several ornamental *Cyperus* have been reported as naturalised at times in warmer areas. Papyrus (*C. papyrus*) is a tall, triangular-stemmed plant to 3 metres and more in ideal conditions, topped with flowerheads of slender filaments, and is the dominant plant in many tropical wetlands within its natural range. Dwarf papyrus (*C. prolifer*) is similar in general appearance, though much more compact in its growing habit, reaching around 1 metre with a more compact, bristly flowerhead. The ornamental umbrella sedge (*C. alternifolius*) is occasionally reported as naturalised though there is little sign that it is likely to become an environmental weed, while a further ornamental species *C. percamenthus* has only recently appeared on the lists of one nursery. The small, annual species *C. tenellus* may be a minor weed on disturbed sites and in created wetlands for several years, until other wetland plants establish and shade it out.

Origins: papyrus is primarily an African plant but is probably also native to southern Europe, while dwarf papyrus is from southern Africa. Papyrus is the most widespread species of these two and is reported as naturalised in both eastern and western Australia, while dwarf papyrus (*C. prolifer*) may sometimes be found naturalised in urban areas in Queensland and New South Wales.

Uses: primarily grown as ornamental water garden plants. Papyrus was the plant from which paper was made in dynastic Egypt, and its rhizomes are reported to be edible (though they are hardly a gourmet food).

Preferred growth conditions: nutrient rich soils in warmer climates, with papyrus growing in deeper waters (1 metre and more) than *C. prolifer.*

Confusing species: none.

Environmental effects: the established ornamental species are minor and localised weeds, but nothing is known yet of the weed potential of *C. percamenthus.* The prolific growth of true papyrus in ideal conditions suggests it may have wider potential as a weed, but there is little sign that it will become one in Australian conditions even though it has been widely available and planted for at least a century.

Control and management: not necessary to date, and an excavator or backhoe is likely to be able to deal with any naturalised patches on the scale reported so far.

Manual removal of seedlings once a year for the next two to three years should complete eradication in most situations.

Eleocharis (Cyperaceae)

The stems of these leafless sedges are tipped with a cone-like spikelet, hence the common name of spikerush. They vary widely in size from relative giants such as the indigenous tall spikerush (*E. sphacelata*, discussed earlier) to low mats of threadlike leaves rarely more than a few centimetres high.

Origins: Variable spikerush (*E. minuta*) from sandy lake shores of Africa and Madagascar is patchily naturalised in eastern Australia, *E. pachycarpa* from Chile is reported from Centennial Park, Sydney, and *E. parodii* from Argentina is recorded as a rice field weed in the Griffith area, New South Wales.

Preferred growth conditions: wetland fringes, apparently only on disturbed ground where there is little competing vegetation.

Confusing species: many indigenous spikerush species. Accurate identification is essential before any control measures for an unidentified spikerush should be considered.

Environmental effects: minor and localised weeds at the present time.

Control and management: all introduced spikerushes are potential environmental weeds and should ideally be eradicated wherever they are found, but they have only been considered a problem in rice fields to date.

Isolepis (Cyperaceae)

Budding clubrush (*I. prolifera*; see plate 8) is a slender-leaved, sprawling tussock that can spread rapidly from new plants formed where the terminal flowerheads come in contact with wet soil or water, a habit known as proliferation. Various other smaller, mostly annual *Isolepis* species are also weedy in a minor way including the tiny *I. hystrix* which is widespread in south-eastern Australia, while at the other extreme *I. sepulcralis* is only known from fairly localised escapes from nurseries. *I. marginata* and *I. cernua* may be native or may include introduced forms, but in any case none of these smaller plants are regarded as serious environmental weeds and they are not likely to be common in undisturbed situations.

Origins: budding clubrush is from southern Africa, now naturalised from garden escapes in parts of eastern New South Wales and south-eastern Western Australia, and more patchily in Victoria.

Preferred growth conditions: shallow waters and wet places in full sunlight, but also able to grow within mats of other aquatic or floating plants, from which it can be difficult to remove completely.

Confusing species: budding clubrush looks much like a larger and coarser version of the widespread native *I. inundata* (which also spreads by proliferation), but has a larger, denser and heavier cluster of flowerheads with up to 25 separate spikes.

Environmental effects: regarded as a moderately serious weed because of its potentially rapid spread, and certainly an undesirable and fairly conspicuous species that should be removed wherever possible. However, in many situations it only grows patchily among indigenous plants, and as it is not apparently toxic may have little impact on the habitat values of natural wetlands.

Control and management: exposed plants are shallow rooted and not difficult to remove by hand, but it is difficult to remove all fragments of the flowerheads, which may be inconspicuously rooted in soil. Seed may also germinate in clear shallow waters or on disturbed, waterlogged soils for two to three years after mature plants are removed.

Juncus (Juncaceae)

Rushes are often found in or at the fringes of wetlands and some native species can be regarded as weeds in seasonally wet pasturelands, as these are among the few indigenous plants that survive grazing and trampling by cattle. Although these colonising species may be much more abundant in disturbed or cleared places than they were before European settlement, they are not considered here as weeds.

Spiny rush (*J. acutus*)

Spiny rush (see plate 10) forms large, open tussocks of distinctly dark, sometimes blackish-looking stems to 2 metres high, each tipped with an evil spike that is likely to break off under the skin on contact.

Origins: apparently native in both Europe and northern America, extending southwards into Africa, and now widely naturalised in southern Australia.

Preferred growth conditions: seasonally wet, often saline soils where there are few competing plants, sometimes in places which may flood at times. Unlike the comparably dark-looking, also spiny-tipped but indigenous sea rush (*J. kraussii*), spiny rush will not thrive in regularly flooded situations.

Confusing species: spiny rush can be confused with sea rush but the much larger and more conspicuous seed capsules are from 4 mm to 6 mm long, while those of sea rush are only around half this length.

Environmental effects: the spiky growth habit may provide some shelter from hawks for small animals, but spiny rush is otherwise avoided by all larger creatures. This is a serious environmental weed that has long been declared noxious in some places, and should ideally be tackled on a national scale.

Control and management: fresh seed is prolifically produced and germinates readily on disturbed ground, even on saline clay soils, but is so small that most of it is only likely to survive for two to three years. The small size of the seed also means it is much less likely to establish where competing plants are already established, so even a shallow layer of mulch will inhibit germination, as will overplanting with shade-generating indigenous species. Any seedlings that do appear should be removed as soon as they are noticed, and at smaller sizes are not difficult to pull up with a gloved hand, using a small, broad-bladed mattock in the other hand to provide a little leverage on the relatively shallow rhizome. Larger tussocks that are on accessible ground can be slashed to ground level during drier times of the year. Continuing this practice annually as a short-term measure will also prevent seed formation, while forcing the deeper rhizomes to grow towards the surface where they can be more easily dealt with later. Many infestations of this species are on exposed, sloping ground which may be flooded at times, so no attempt to eradicate it completely should be made until other plants are already established enough to prevent erosion.

Other introduced *Juncus* species

Several introduced species of relatively soft-tissued, hollow-stemmed rushes are established weeds in parts of southern Australia, the most widespread being jointed rush (*J. articulatus*; see plate 10). Small-headed rush (*J. microcephalus*) is less common but is believed to be expanding in range. Soft rush (*J. effusus*) is a proven weed in many places but is sometimes sold by nurseries as corkscrew rush (cultivar 'Spiralis'), a mutant form with bizarre coiled stems resembling broken springs. Although this botanical oddity may seem to be too unnatural a form to survive outside of water gardens, it sets seed freely with around 50 per cent of the seedlings reverting to the weedy, wild type. Other introduced species such as *J. oxycarpus* have been recorded as locally common but are very restricted in range.

Origins: jointed rush and soft rush are believed to be native to temperate Europe and Asia, northern America and even as far south as Africa, while small-headed rush is a South American import.

Preferred growth conditions: all of these plants thrive in seasonally wet or shallowly flooded places, but jointed rush grows and spreads most vigorously in permanent, shallow waters up to around 30 cm deep. In ideal conditions or on rich, flooded soils these various species may reach between 50 cm and 1 metre in height, but in more ephemeral waters or where they are regularly cut they are more likely to be smaller, growing as a carpet of more slender and crowded stems.

Confusing species: introduced rushes are not always easy to separate from a number of related indigenous species with comparable growth habits, particularly as all of these plants vary in appearance under differing conditions of shade and seasonal flooding. Accurate identification requires close examination of the seed capsules and their surrounding tepals, and the arrangement of the flowerheads on the flowering stalks. If control measures are being considered for any apparently weedy species it is essential to have identifications confirmed by a herbarium or qualified botanist.

Environmental effects: these are primarily weeds of cleared ground, growing reasonably well even on raw clays, though they may also invade the fringes of natural wetlands. Given the wide range most of these introduced rushes have already occupied, their general similarity in appearance to many indigenous rushes, and that they are unlikely to be recognised as weeds until they are well-entrenched enough to become obvious, we can only hope that they are also ecologically comparable to their indigenous relatives because there isn't much prospect of getting rid of them. Instead, all efforts should be focused on minimising the spread of introduced rushes into new places.

Control and management: in the case of jointed rush it is unlikely that there is any purpose to control, as this species is widespread and difficult to identify when not flowering or seeding, and has already probably colonised most available niches within its present range. Although it might be possible to restrict the spread of the less widely distributed weedy *Juncus* species, given that there is no move being made to do so at either federal or state levels and that there are many far more urgent weed problems to be dealt with, it is likely that these will also soon be beyond any real prospect of containment.

Lilaea (Juncaginaceae)

Lilaea (*L. scilloides*) is an annual weed of shallow, seasonal waters, with the spongy, cylindrical leaves becoming more flattened closer to the tip.

Origins: originally from the Americas, now also patchily naturalised inland in rice-growing areas of Victoria and New South Wales and occasionally elsewhere.

Preferred growth conditions: shallow seasonal waters, even temporary pools with low nutrient content.

Confusing species: could be confused with small plants of some indigenous *Triglochin* species.

Environmental effects: none known at the present time, though it is briefly treated as a potential environmental weed in Adair *et al.* (2008).

Control and management: manual removal of this shallow-rooted plant before flowering should give complete control, as the seed is likely to be short-lived.

Schoenoplectus (Cyperaceae)

The tall and leafless Californian clubrush (*S. californicus*) is sporadically naturalised in New South Wales, though an excusable confusion with river clubrush (*S. tabernaemontani*, a widespread native) may be masking the actual extent of its spread. The much smaller species *S. erectus* and *S. lineolatus* have also been recorded as naturalised in eastern Australia, but are not considered here as they are very localised and uncommon, and are unlikely to invade relatively undisturbed wetlands on present indications.

Origins: Californian clubrush is widespread along the Pacific side of the Americas.

Uses: harvested for starch from its rhizomes, while the stems have also been used for many purposes from boat-building to construction. Native relatives of this potential weed are likely to be equivalent in every way, if only local research was carried out on appropriate timing of harvests for the diverse purposes this plant has been used for. It has long been clear that harvesting time (as guided by local knowledge) is perhaps the single most critical factor in effective use of many types of sedges worldwide, including this species.

Preferred growth conditions: probably comparable to river clubrush, preferring reasonably deep, permanent waters, sometimes even tidal and slightly saline.

Confusing species: river clubrush and, to a lesser degree, the more tropical and salt-tolerant *S. litoralis* which is also indigenous.

Environmental effects: none at the present time, but this species is a potential competitor for the native river clubrush. If allowed to spread further, California clubrush could create a comparable situation to the cat-tail versus cumbungi (*Typha*) problem, to be discussed shortly, where very few Australians are aware that a fast-spreading introduced species may be replacing the indigenous species until it is too late to do much about it.

Control and management: ideally, the relatively few and localised populations of this potential weed should be eradicated before they have the chance to spread.

Sparganium (Sparganiaceae)

Branching burr-reed (*S. erectum*; see plate 7) is a fleshy-leaved, upright plant with sharply triangular leaf-bases. The spherical flowerheads are made up of soft-spined seeds supported by zig-zag flowerstalks, the branching of which has been used to help with identification of species in the northern hemisphere, where most species are found.

Origins: forms of this plant are found from Africa to Europe, and probably into western Asia. Victorian populations were long regarded as indigenous, though some records suggest it may have been introduced with sheep from the earliest agricultural settlement at Portland, and its recent proliferation in parts of south-eastern Queensland suggests that this is a weed which is only just reaching regions and climates more suited to its spread.

Preferred growth conditions: still or slow-moving waters to around 1 metre deep, tolerating deep shade and spreading by runners under such conditions, but in southern Australia apparently only flowering and setting seed in sunlit places.

Confusing species: the indigenous floating burr-reed (*S. subglobosum*) is a smaller and narrower-leaved relative, found from eastern Victoria northwards to Asia. Surrey Jacobs has expressed doubts (personal communication) that the relatively large plants under this name in Victoria are necessarily the same as the distinctively slender, more northern forms, and it may be worth reassessing the relationships of these various plants.

Environmental effects: at present a relatively minor and localised weed in southern Australia, but a burst of new records from southern Queensland beginning in the 1960s indicates potential for further spread.

Control and management: the environmental impacts of branching burr-reed are not known, and my impression is that it mainly grows in places where ecologically competitive natives don't thrive – including in the dense shade of willows. Given the presently localised distribution of this species, and the ease with which it can be pulled out of soft mud in relatively deep waters, many populations could be largely eradicated by one or two people in wetsuits throwing the lifted plants well above water level, with a follow-up treatment to complete the job a couple of years later.

Typha (Typhaceae)

Cat-tail or lesser reedmace (*T. latifolia*; see plate 3), along with its two indigenous relatives narrowleaf cumbungi (*T. domingensis*) and broadleaf cumbungi (*T. orientalis*), has been a part of many wetland landscapes in south-eastern Australia for a whole generation, and few people seem aware that it is actually a wide-ranging and invasive weed. Although it is illegal to import other exotic *Typha* species, two less invasive exotic species are also sold by water garden nurseries. The weed potential of *T. laxmannii* is unknown but no naturalised populations are recorded even though it has been available for decades, while the circum-Arctic *T. minima* rarely flowers north of Tasmania, and may not survive a hot summer's day. The distinctive poker heads of all *Typha* species produce huge amounts of tiny seeds with a fluffy 'parachute' that allows them to travel considerable distances in the wind.

Origins: cat-tail was introduced as an ornamental plant from the northern hemisphere, and for the past few decades has been spreading rapidly in Victoria and Tasmania, appearing even in some apparently pristine situations. The two indigenous species are found throughout Australia, but broadleaf cumbungi is regarded as introduced in Western Australia, and is now patchily established around the Perth area.

Uses: all *Typha* species produce shoots with an edible core, and edible starch can be extracted from the rhizomes. Pollen is shed by the short-lived male poker which soon disintegrates, leaving only a faint scar above the more familiar female poker below, and has been used to make bread by indigenous peoples worldwide, though it is now often mixed with flour for binding purposes.

Preferred growth conditions: mature plants of all three species will grow to 1 metre or more deep, but seed mainly germinates on waterlogged soils where there is little or no shading or competing vegetation. Seedlings are slow growing on newly exposed soils, but their growth rate accelerates as nutrient-rich silt accumulates. The ultimate size of the mature plants is limited by availability of nutrients and water, and although cumbungi may spread along drains, the plants remain relatively stunted if the channels dry out with any regularity.

Confusing species: cat-tail has a dark, chocolate-brown female poker which often touches the short-lived male poker above, with little or no gap between them, and its foliage is blue-green rather than the yellow-green of the indigenous species. The two native species (see plate 2) have paler, tan-brown pokers, usually with a distinct space between male and female flowerheads, and are not always easy to tell apart. Narrowleaf cumbungi is a smaller and more slender plant to around 2 metres high, with narrow pokers which may be 40 times as long as they are wide. Broadleaf cumbungi is a more robust plant reaching around 3 metres high, with proportionately thicker and often much larger pokers. However, there is some overlap in poker size and shape between these species, so intermediate plants can only be separated by microscopic examination of the flowers.

Environmental effects: all *Typha* are colonising plants that may be more abundant now than before European settlement, as they are particularly abundant in open drains and other disturbed or exposed situations including farm dams, where they may adversely affect water quality and limit stock access. In natural wetlands cumbungis provide a limited type of habitat, which is only used by a few ecologically adaptable species such as swamp hens.

Control and management: mature stands of *Typha* species may spread indefinitely on rich soils, as long as the water depth isn't much greater than 1.2 metres, but excavation of deeper channels will usually limit their spread. There is little prospect of eradicating well-established populations unless the main job of clearing can be done from shore by a long-reach excavator, and any remaining plants can be pulled out by wet-suited workers a year or so later. The tiny seed of *Typha* species is so small it won't germinate readily among other plants, and new seedlings are shaded out by almost any other wetland species, so preventative plantings may be all that is required to prevent *Typha* seed from establishing in a new location.

At a more specific level, the marked similarity of cat-tail to the cumbungis has meant that has not been perceived as a serious threat, and it may now be too late to control the spread of this introduced weed. For example, a small patch of cat-tail in a small town several kilometres from where I live was ignored by the local council even after I pointed it out 15 years ago. Yet this was just the tip of an ever-expanding iceberg, and I have recently found a much older and well-entrenched population accessible only by canoe, which probably arrived in the headwaters of a nearby natural lake formed by a landslide more than 50 years ago. We can only hope that cat-tail is ecologically comparable to our native species, and no less benign.

Other wetland weeds

Potentially serious aquatic weeds such as *Stratiotes aloides* and the various species of *Trapa* are not included here because they have not been introduced into Australia, and hopefully never will be. The sole record of *Trapa* in Hibbert (2004) is a case of confusion with the very different water chestnut (*Eleocharis dulcis*), which the nursery in question has yet to acknowledge.

Alisma (Alismataceae)

Of the two species of this genus in Australia one (see plate 30) is usually regarded as indigenous, while the other is an introduced weed. Both are native to western Asia, Europe and northern Africa, but water plantain is also accepted as native from south-western Victoria to the eastern half of New South Wales. A third possible species has recently been offered by a water garden nursery.

Water plantain (*A. plantago-aquatica*)

A tall herb with broad leaves carried on metre-high stems, and open masses of tiny flowers carried on stems to 1.5 metres. Although generally accepted as native and therefore only regarded as a weed in rice fields, water plantain can also be abundant in disturbed situations in wetlands, in open drains, and on seasonally flooded pastureland.

Origins: widespread in Eurasia and northern Africa, and while the relatively restricted range in south-eastern Australia seems unusual for an adaptable and long-established plant, it is broadly comparable to that of *Lythrum salicaria*, discussed in Chapter 3.

Uses: possibly of some significance as habitat, otherwise occasionally offered as an ornamental indigenous plant.

Preferred growth conditions: grows most abundantly on disturbed ground that may be seasonally flooded, the seed germinating prolifically as water levels fall. Seed viability is reputed to be relatively short, though some may still germinate after three years of dry storage.

Confusing species: both water plantains could be confused with some *Sagittaria* species when not in flower. The two water plantain species can be readily

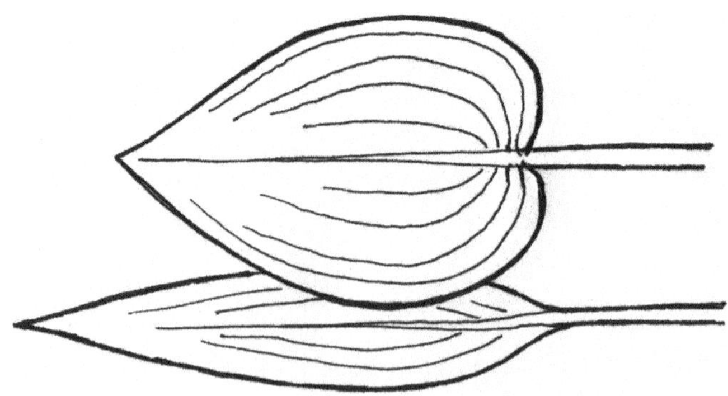

Alisma species: Mature leaves of *A. plantago-aquatica* above, with *A. lanceolata* below; immature leaves of the two species look very similar.

separated from each other when fully mature by height and leaf shape, though young plants with floating leaves can be confused with *Damasonium* (discussed earlier) and even with the native swamp lily (*Ottelia ovalifolia*), although the floating leaves of the latter species are never held above water.

Environmental impacts and values: pure stands of water plantain in some places suggest it may be allelopathic to some degree, though these may also be the result of repeated disturbance which inhibits the growth of slower-growing competing species. Specific habitat values are not known and I have rarely found indigenous animals among this plant, but it may grow densely enough to be a significant shelter plant.

Control and management: regardless of origin, water plantain doesn't seem to be moving beyond its long-established range in south-eastern Australia of its own accord. As there are many other areas in southern Australia where it could potentially establish, and it can spread prolifically from seed within a year or two of establishment, this species should be banned from sale beyond its present Australian range. Deliberate planting in wetlands except in areas where significant populations are already well-established should be discouraged.

Narrowleaf water plantain (*A. lanceolata*)

A smaller water plantain species with leaves that only reach around half the height of water plantain, remaining relatively narrow and elongated even on mature plants, at present this weed is relatively localised from around Rushworth in north-central Victoria to south-central New South Wales. A further potentially weedy species has been offered through the water garden trade in recent years as *A. parviflora*, which is nominally a variety of *A. plantago-aquatica*, but the photos posted on the internet look suspiciously like *A. lanceolata*.

Preferred growth conditions: grows most abundantly on disturbed ground that may be seasonally flooded, with the seed germinating prolifically as water levels fall.

Confusing species: see comments under *A. plantago-aquatica*, above.

Environmental impacts and values: narrowleaf water plantain is mainly a weed of rice fields and disturbed habitats such as drains at the present time, with unknown weed potential in less disturbed habitats. It may become a more serious weed with time and has the potential to spread downstream through many hundreds of kilometres of the Murray River catchment.

Control and management: it may still be possible to eradicate narrowleaf water plantain if a concerted push was made to do so over its restricted range, and any nursery-grown plants under this or any other name are eradicated. The seed is reputed to be fairly short-lived, and germinates rapidly in suitable conditions, so an intensive, focused effort over one growing season may be enough to achieve local extinction – especially in recent years since greatly reduced water quotas mean that rice fields are only likely to be flooded once in a few years. Although there could be some collateral damage to young plants of *A. plantago-aquatica* in the course of dealing with *A. lanceolata*, the putatively native species is much more abundant, and demonstrably more resilient than the proven weed.

Alternanthera (Amaranthaceae)

Alligator weed (*A. philoxeroides*; see plates 1 and 28) forms extensive, rooted mats in shallow waters and can float out considerable distances into slow-moving waters; it also grows in a more stunted form or even as a creeping carpet in drier situations, such as lawns where rainfall is reliable enough to sustain it. It has been declared noxious or otherwise prohibited in most parts of Australia.

Origins: native to tropical and subtropical South America, but now regarded as a serious weed in some 30 countries. Within Australia it was localised primarily around the Sydney to Newcastle region as well as in several other New South Wales localities until the 1990s, but is now found in urban areas from Melbourne to Brisbane.

Uses: deliberately spread as an edible plant due to confusion with *A. sessilis* and other edible relatives from South-East Asia, and at one stage widely sold in Asian markets and even by mail-order.

Preferred growth conditions: most vigorous growth is in tropical to subtropical areas in shallow, warm waters, but it can still be very invasive even in cooler, more southerly areas as long as enough nutrient is available. Plants will even grow into well-watered crops and turf in a reduced form which may be difficult to see, and translocations of such turf have probably helped spread this species within New South Wales. Severe frosts will kill emergent parts of the plant, but do not affect the submerged parts that regrow in warmer conditions.

Confusing species: several native species of *Alternanthera* (particularly *A. angustifolia* and *A. sessilis*), though in native species the flowerheads are attached directly to the main stem at the base of the leaves, rather than held on a stalk up to 9 cm long (and also originating from the leaf bases).

Environmental effects: a fast-spreading and adaptable weed that thrives in both water and on land, that may sometimes creep into relatively undisturbed wetlands. As it often grows in monospecific stands, it also restricts the types and diversity of habitats available for native animals, and although some smaller fishes will shelter in it they are more likely to be the introduced plague minnow (*Gambusia holbrooki*) than any native species. New plants will grow from surprisingly small fragments, but little or no seed appears to be produced under Australian conditions.

Control and management: this weed shows considerable resistance to a range of herbicides, and in most cases is not only likely to recover better from frequent herbicide treatments than the surrounding wetland habitats, but may fragment into viable pieces more readily as a result of such treatment. Small infestations growing in water can be fairly effectively removed by hand, but care must be taken to collect any fragments that drift away. Biological control appears to be the best long-term prospect for managing alligator weed, and although there is the problem that such control agents may affect indigenous *Alternanthera* species, the indigenous plants may themselves be a potential source of biological controls (including diseases) that alligator weed has yet to come in contact with. In the longer term, and assuming that alligator weed does not develop the capacity to produce viable seed, it will hopefully reduce in vigour as viruses and other internal problems proliferate.

Alligator weed has the potential to invade many kinds of wetlands from the tropics to Tasmania, and even the many recent infestations as far south as Melbourne may already be beyond any real prospect of control. Control of this species is regarded as an environmental priority Australia-wide, and is being tackled at this level, so any new or suspected infestations should be reported to state authorities immediately. An *Alligator Weed Control Manual* (van Oosterhout 2007) can be downloaded from the Weeds of National Significance website, and is also available in hardcover.

Annona (Annonaceae)

Pond apple (*A. glabra*; see plate 26) is a medium-sized, semi-deciduous tree closely related to several highly regarded tropical fruits, including custard apples and soursop. A slow-developing weed, it has spread gradually over many decades, and it is only relatively recently that the scale of infestation has been recognised.

Origins: central and South America northwards to Florida, where ironically it is regarded as increasingly endangered due to the widescale drainage of the Everglades. In Australia, the main areas where this weed has entrenched itself are along the tropical Queensland coast, with outlying patches as far south as Brisbane, while small infestations around the Darwin and Arnhem regions have apparently been successfully controlled.

Uses: the fruit is edible but not particularly good, and in horticulture it has mainly been used as a root stock for grafting some of its better-quality relatives, so that these can be grown on seasonally boggy soils.

Preferred growth conditions: wet soils which may dry out seasonally, tolerating shade during the early stages of growth, and some degree of salinity.

Confusing species: when not fruiting or flowering, various broad-leaved mangrove species that may also be found in the upper, fresher reaches of mangrove swamps.

Environmental effects: although there are only an estimated 2000 hectares of pond apple infestation in Australia at the present time, and it grows most readily in cleared and disturbed habitats, this species will also colonise clearings in the fresher reaches of mangrove swamps. The relatively large seed floats and is dispersed by water, surviving seawater for weeks so it may end up in inaccessible upper mangrove areas hundreds of kilometres away. It is not clear whether the increasing rate of infestation is due to an innate ability to establish among other vegetation, or if the plant is taking advantage of other changes occurring within many of these places – such as the proliferation of feral pigs, which spread the seed and disturb soil with their rooting activities. Seed may also be dispersed by larger fruit-eating birds such as cassowaries. Regardless of cause, pond apple is already an established environmental weed in a number of Queensland's National Parks.

Control and management: cutting individual trunks and painting the fresh cut with relatively concentrated herbicide is reported to be highly effective, the main difficulties being access to many sites, and the ever-increasing number and size of protected saltwater crocodiles in many of these areas. Biological control is not being considered for pond apple as its close relatives in Australia are commercially valuable species. See the Pond Apple National Strategic Plan on the Weeds of National Significance website for (reasonably) current updates what is being done about this weed.

Aponogeton (Aponogetonaceae)

Water hawthorn or cape pondlily (*A. distachyos*; see plate 19) produces elongated, floating leaves and two-tiered flowerheads held above the water's surface. It has long been sold as an ornamental plant because of its autumn to spring flowering habit, which complements the flowering season of waterlilies (*Nymphaea*).

Origins: southern Africa, now also patchily naturalised through southern Australia.

Uses: cultivated in South Africa for the fragrant, edible flowers; larger tubers produce a fine-grained starch comparable to arrowroot.

Preferred growth conditions: moderately deep and often slow-flowing waters up to 1 metre deep, in situations which may dry out completely at times, and regrowing from the dormant tubers after autumn rains.

Confusing species: the leaves are very similar to the indigenous swamp lily (*Ottelia ovalifolia*), but the three large, well-opened petals of swamp lily are very different in appearance from the upright columns of water hawthorn. In more northern areas (where water hawthorn is rare) it may be confused with some native *Aponogeton* species with floating leaves, though these have much more slender and less conspicuous flowerheads.

Environmental effects: I have generally regarded this plant as a minor environmental weed, as it has reportedly been naturalised in southern Australia for around 150 years, yet most populations only occupy small sections of streams even where they have long been established. However, over the course of a recent prolonged drought in southern Victoria, some populations such as near Port Campbell have become far more prolific, so this plant may become increasingly weedy with global warming.

Control and management: the short-lived fresh seed sinks and germinates almost immediately, so young plants don't appear far from their parents. In soft silt, the tubers can be dug out with a gloved hand, an effective control for smaller populations as most of the remaining seedlings can be easily located and removed in the following autumn.

Aristea (Iridaceae)

A. ecklonii (see plate 21) is an uninspiring relative of the irises, with small, deep-blue flowers that produce considerable volumes of seed. Although this potential weed is no more attractive than several of its Australian relatives, particularly *Orthrosanthus* and other native flags, it is used as a landscape plant as well as in gardens.

Origins: from southern to tropical western Africa, naturalised in Victoria and New South Wales.

Uses: a minor ornamental.

Preferred growth conditions: an adaptable plant, growing in any area with reasonably reliable rainfall and tolerating waterlogging for long periods.

Confusing species: could be confused with several Australian relatives.

Environmental effects: not known.

Control and management: the large seeds don't spread far from the parent, but germinate freely over many years so that hundreds of seedlings may appear within a few metres of any clump which has flowered. The shallow-rooted tussocks can be removed by hand, and as there is no deep rhizome this will destroy all established plants. However, a watch should be kept for germinating seedlings for up to another five years.

Aster (Asteraceae)

Bushy starwort (*A. subulatus*) is a fast-growing, upright, annual daisy to 1.5 metres tall, found on disturbed or open soils that may be shallowly flooded at times.

Origins: originally from northern America, now also naturalised across much of non-tropical Australia.

Preferred growth conditions: seasonally wet, exposed soils with relatively few competing plants, and preferring some degree of salinity.

Confusing species: with its upright habit, small daisy flowers and masses of fluffy seed, bushy starwort is unlikely to be confused with other plants found around the fringes of slightly saline wetlands.

Environmental effects: a common weed that has become even more abundant over the past decade in parts of south-eastern Australia, as it colonised newly

Aster subulatus.

exposed lake shores over the course of years of drought. Although of no value as habitat for any native animals, even including butterflies and moths, it generally grows sparsely enough that it does not provide obvious competition for indigenous plants in such habitats.

Control and management: resistant to many herbicides, fast-growing and spreading by countless windblown seeds, bushy starwort has probably entrenched itself everywhere its seed is likely to be able to reach. Control is likely to be impossible, except on a small, regional and temporary scale, assuming there is any real reason to put the effort into it. Although tolerating seasonal flooding, bushy starwort can be drowned after a few months if water levels can be raised for long enough.

Bacopa (Scrophulariaceae)

Bacopa (*B. carolineana*) is a round-leaved, blue-flowered aquarium plant that has naturalised in the Sydney region.

Origins: south-eastern USA.

Uses: introduced as an ornamental pond and aquarium plant.

Preferred growth conditions: warm, shallow waters, and tolerant of some shading.

Confusing species: the indigenous but smaller-leaved and usually paler-flowered native, *B. monniera*. Other species and forms of bacopa are also grown for the aquarium trade but have not yet been recorded as naturalised.

Environmental effects: a minor and localised weed only at this stage.

Control and management: physical removal is effective for small infestations.

Betula (Betulaceae)

River birch (*B. nigra*) is a tree with dark, scraggy, peeling bark, and can ultimately reach 30 metres in height.

Origins: North America, reported as naturalised around some wetlands in Western Australia.

Uses: a not particularly ornamental tree sometimes recommended for planting in permanently moist to wet ground, with a number of named cultivars available, though given its ultimate size not suitable for most garden situations.

Preferred growth conditions: rich soils along streams and lakes.

Confusing species: none.

Environmental effects: a relatively minor weed at present.

Control and management: although black birch is unlikely to be as weedy as willows or poplars, if it continues to appear as a weed in riparian situations restrictions on its planting should be considered.

Cabomba (Cabombaceae)

Fanwort (*C. caroliniana*; see plates 1 and 22) is related to the indigenous water shield (*Brasenia schreberi*), but with tiny, floating leaves that support the smallish, white-petalled flowers. The finely divided underwater foliage is much more conspicuous, forming dense, submerged tangles even in quite deep and shaded places.

Origins: native in subtropical South America but originally described from south-eastern USA; the considerable distance between these disjunct locations has led some botanists to suggest it was introduced to the USA. Fanwort is now a weed in some Asian countries, and is patchily naturalised along the eastern coast of Australia from central Victoria to tropical coastal areas of Queensland, as well as near Darwin.

Uses: an ornamental plant now banned through most of Australia. Many of the known populations have almost certainly been deliberately naturalised by aquarium suppliers, and during the early 1990s one aquarium plant wholesaler was actively contacting water garden nurseries to encourage them to grow or distribute this weed, with guarantees of a steady market for its sale.

Preferred growth conditions: reasonably clear waters up to at least 3 metres deep, also tolerating quite dense shade for months at a time.

Confusing species: could be confused with hornwort (*Ceratophyllum*) although this is much coarser- and stiffer-leaved, and does not produce roots. Submerged growth of some indigenous water milfoils (*Myriophyllum*) may look superficially similar, but in water milfoils the submerged leaves are toothed with many prongs forming a two-sided comb along a central rib, rather than forking repeatedly as in fanwort.

Environmental effects: an aggressively competitive weed that can displace or overrun many indigenous submerged aquatics, perhaps even the weedy *Hydrilla*, growing densely enough that it will significantly reduce the free passage of larger native animals from fishes to water rats. Fanwort is also regarded as a serious potential threat to water supplies, as it grows in deep waters over a wide range of climates from Tasmania to the tropics, producing a distinctive, unpleasant taint if present in large quantities. Where large masses of fanwort are present, the accumulation of dead and dying plant material lowers dissolved oxygen levels, making it even more of an environmental liability.

Control and management: larger infestations are almost impossible to control and may spread downstream from broken fragments in the right conditions, but it is suspected that all naturalised populations in Australia are the result of deliberate introductions. Plants can be killed by drawdown in reservoirs if the soil it grows on is exposed to the air for several weeks, a successful treatment at Palmerston in the Northern Territory though the cost was considerable. On the positive side – there is no evidence of seed being formed in Australia, and seed viability is reported to be only around two years even in wild populations. It is possible that in the long term this species will die out as viral and other disease problems reduce its vigour.

Callitriche (Callitrichaceae)

The cosmopolitan common starwort (*C. stagnalis*; see plate 27) and to a lesser degree thread starwort (*C. brutia*) are common and often vigorous aquatic weeds, forming dense, floating mats of rounded leaves. Several other introduced species are also recorded in parts of southern Australia, of which *C. palustris* may be indigenous, but these uncommon, usually localised species are not discussed here.

Origins: the two weedy introduced starworts are native to Europe and western Asia, with common starwort also found southwards into Africa. In Australia, common starwort is widely naturalised south of the tropics, while the more cold-loving thread starwort is mainly found in the temperate south-east of the mainland.

Preferred growth conditions: clear, slow-moving or still waters up to 1 metre, rooted in nutrient-rich, silty mud. If water levels fall, the stranded plants will grow as a low carpet unless killed off by heat, to be replaced by a copious supply of seedlings when the area is flooded again.

Confusing species: many of the native starworts look similar to the introduced species, although they are mostly less aquatic and are more likely to be found in ephemeral pools rather than in more permanent waters.

Environmental effects: among the ranks of introduced aquatic weeds starworts are relatively benign, even useful in some situations, which is fortunate as there is little chance of ever getting rid of them. Plants may be eaten by some herbivorous waterbirds, while the dense mats trailing in slow streams provide shelter for shrimps and smaller fishes, and in nutrient-rich, disturbed situations dense mats of starwort may prevent outbreaks of cyanobacteria (so-called blue-green algae).

Control and management: small infestations are easily pulled out by hand before they set seed, but it is important to tease every fine thread of stem out of the mud. Where starworts have been long established, disturbance is only likely to trigger dormant seed into growth.

Crassula (Crassulaceae)

C. natans var. *minus* forms a sprawling, reddish mat on low-lying ground that may be inundated for part of the year, trailing along the water's surface when submerged.

Origins: southern Africa, now naturalised in south-eastern and south-western Australia.

Preferred growth conditions: moist, shallow soils where there is little competition from taller, shading plants.

Confusing species: could possibly be misidentified as a small variant of the native swamp stonecrop (*C. helmsii*) when flooded.

Environmental effects: a minor weed, patchily distributed in places where it is not overshaded by competing species.

Control and management: removal of small patches can be done manually. It is probably not worth dealing with more widespread infestations unless this species begins to show more aggressive tendencies than it has to date, or is recorded in any significant numbers from undisturbed wetlands.

Echinodorus (Alismataceae)

Swordplants are mainly used in aquaria as specimen plants, but some smaller species are also grown as an underwater lawn.

Origins: most swordplants are from tropical and subtropical South America, with several extending to the southern USA.

Uses: ornamental aquarium plants.

Preferred growth conditions: variable in climate tolerances depending on species, with the radicans sword (*E. cordifolius*) becoming winter dormant so that it can survive regular light frosts, and the melon sword (*E. osiris*) thriving underwater through southern winters as long as the water's surface does not freeze.

Confusing species: some other plants of the same family, particularly *Alisma* and *Sagittaria* species.

Environmental impacts and values: most swordplants sold at the present time are hybrids or selections with little or no weed potential. However, some of the older selections and species could potentially naturalise in subtropical areas, or even further south as global warming continues, and *E. cordifolius* was abundant in the ornamental lake at Brisbane Botanic Gardens in the late 1990s. An unnamed species of 'dwarf' swordplant tentatively identified as *E. latifolius* has thrived and spread rapidly in my nursery when grown as an emergent plant, surviving even cold winters.

Control and management: an increasing emphasis on more colourful hybrid swordplants means that the few potentially weedy species or forms are now rarely available. Further imports of species of swordplants should be assessed more carefully than they have been in the past, and all species and varieties offered for sale should be accurately identified so that any potential as weeds can be assessed for every clone grown in Australia.

Egeria (Hydrocharitaceae)

See discussion under *Elodea*.

Eichhornia (Pontederiaceae)

Water hyacinth (*E. crassipes*; see plate 17) is a large, floating, tropical plant with swollen stems supporting blunt, kidney-shaped leaves; the younger plants are more compact and globular. This species has been widely introduced as an ornamental for its upright stems of lilac flowers, spreading rapidly in warm conditions by runners to form a dense mat. It is regarded as a serious weed almost everywhere, including Australia where it is a prohibited plant at the national level.

Origins: tropical South America, now widely naturalised in the tropics, and growing vigorously even as far south as Victoria.

Uses: the soft fibrous stems have been used to make woven furnishings in recent years, and plants have been eaten as a famine food. It has also been opportunistically used in water treatment, methane production and as stock feed, but there is no country in the world where water hyacinth has been introduced that would not be better off without it.

Preferred growth conditions: warm, slow-moving or still waters preferably with a high nutrient content, but also able to grow on mud. Although primarily a tropical species, it will survive regular light frosts, recovering rapidly with the onset of warmer conditions in spring.

Confusing species: mature plants of water hyacinth are among the most distinctive aquatic plants, but at younger stages may be confused with the rooted, submerged *E. azurea*, which was being sold by some water garden nurseries as the native *Monochoria cyanea* in the 1990s. However, both of these grow anchored in shallow waters, and neither develops the distinctive swollen stems of water hyacinth.

Environmental effects: choking mats of water hyacinth change underwater ecology dramatically, cutting off light so no aquatic plants can grow, and reducing oxygen levels so few animals other than mosquito larvae, which can breathe air directly, survive. Water hyacinth can also increase evaporation rates up to five times by transpiration. The history of its horrifically fast spread, and major environmental impacts on fisheries, agriculture and water quality worldwide is summarised in some detail in Parsons and Cuthbertson (2001).

Control and management: control should not be attempted by individuals, except where only a small number of recently arrived or localised plants are involved; even then if seed has been formed, seedlings may continue to appear spontaneously up to 15 years later. Any new or suspected infestations should be reported to state authorities, but there is little chance of controlling the larger and thoroughly entrenched populations in tropical and subtropical Australia, except through biological controls. Weevils and a moth introduced in the 1970s have been reasonably effective at reducing water hyacinth in Australia, along with damage caused by indigenous fungi and bacteria invading the insect-damaged leaves, but the most spectacular results using these species have been in New Guinea and parts of

Africa, where water hyacinth is now apparently in decline in many places. Further south, the emphasis is on preventing further spread, especially from sale plants which still appear from time to time in markets and sometimes even nurseries.

Elodea and *Egeria* (Hydrocharitaceae)

Several related species of similar-looking submerged plants from this single family have been introduced, and have naturalised to varying degrees. Canadian pondweed (*Elodea canadensis*) is by far the most successful, with dense waterweed (*Egeria densa*) more patchily distributed. The relatively soft leaves of these two waterweeds are arranged in whorls around the stems, not in spirals, with small, white, three-petalled flowers that float on the surface (see plate 23). In curled pondweed (*Lagarosiphon major*) the stiff, recurved leaves spiral along the stems, but as this species has only ever naturalised on a localised scale in Australia, and appears to have been successfully eradicated it is not discussed further.

Origins: Canadian pondweed was originally from North America, dense waterweed from South America, and curled pondweed from southern Africa.

Uses: all of these species were introduced as aquarium or pond plants, and dense waterweed continues to be legally sold, increasing the chances that it will naturalise further.

Preferred growth conditions: clear, slow to fairly fast-moving, permanent streams over silt, and in the case of Canadian waterweed even slow-moving irrigation channels.

Confusing species: the two naturalised species can be separated by their distinctive leaf habits from photos alone, with dense waterweed being by far the larger-leaved form. The related water thyme (*Hydrilla verticillata*) is an indigenous relative with distinctively saw-toothed leaf-edges. Although officially regarded as a noxious weed in the Northern Territory, water thyme is a significant habitat plant in many tropical to subtropical wetlands that has disappeared from many more southerly areas, and it is not considered as a weed in this book.

Environmental effects: waterweeds are mainly regarded as a problem in open channels and can form extensive thickets in more permanent waters such as reservoirs. Canadian waterweed is also a common plant in many southern streams, forming dense, submerged mounds. Among introduced weeds these are relatively innocuous plants, used as spawning sites by some native fishes and as shelter by their young, and may act as an ecological replacement for the indigenous *Hydrilla*, which was more widespread before the advent of weirs in the Murray–Darling catchment. The two introduced species are also readily eaten by some herbivorous waterbirds.

Control and management: there are no known means of eradicating these two species once they are established, and even the modest successes reported after drawdowns in reservoirs are unlikely to be long-lasting in their effects. Manual removal is just as likely to encourage their spread, as even small floating fragments will grow. However, early introductions of *Elodea* into England, where this species was distinctly invasive for the first few decades, later became markedly less vigorous and this may also be happening in Australia. It is likely that all plants of the two introduced species in Australia are male clones which means no seed is being produced, and there is some chance that both species will eventually die out as they accumulate an ever-increasing burden of disease organisms, or possibly just of old age.

Epilobium (Onagraceae)

Of the introduced wetland plants in this genus (see plate 29) pinkweed (*E. ciliatum*) is an annual weed of nurseries and disturbed ground, occasionally appearing on the fringes of wetlands but rarely persisting. Tall willowherb (*E. hirsutum*) is a perennial plant reaching around 2 metres. It was apparently introduced as an ornamental garden plant, and became weedy in several sites around the Geelong (Victoria) area before control measures were undertaken.

Origins: pinkweed is from the Americas, its original range probably unknown due to its weedy propensities, while tall willowherb is widespread in the northern hemisphere.

Preferred growth conditions: moist to wet ground that may be flooded at times.

Confusing species: various indigenous species of *Epilobium*; see the earlier discussion of two native species with comparable lifecycles.

Environmental effects: minor weeds only at the present time, but the copious production of fine, wind-blown seed means that either of these species can potentially colonise new ground even hundreds of kilometres away.

Control and management: the few localised outbreaks of tall willowherb have hopefully been contained. While pinkweed continues as a persistent and irritating weed of moist, cultivated soils, it shows little sign of spreading into undisturbed habitats. All of these plants are relatively short-lived and rely on wind to spread their seed, so they can be managed or even eradicated by slashing at strategic times. However, this must be done *well* before flowering begins as even cut stems with green seedpods will continue to ripen seed, as I learned when sending partially dried specimens of tall willowherb to weed expert Kate Blood. These looked very close to death when posted, but to my embarrassment shed copious amounts of drifting seed when the envelope was opened two days later.

Equisetum (Equisetaceae)

Common horsetail (*E. arvense*) is one of a group of primitive plants that remains abundant in parts of the northern hemisphere, and includes a number of species regarded as extremely persistent weeds. The jointed, many-branched stems tipped with globular cones are characteristic. The taller and sparsely branched *E. hyemale* is equally weedy but less water-loving, and the only recorded naturalised population of this species has apparently been destroyed.

Origins: widespread in cooler parts of the northern hemisphere.

Uses: common horsetail has been used by herbalists and also in esoteric preparations purported to increase light absorption by plants, but has been spread most widely among collectors of primitive plants for its well-documented origins in deep time.

Preferred growth conditions: moist to wet soils, tolerating flooding at times.

Confusing species: horsetails could be confused with horsetail restio (*Elegia capensis*) from southern Africa, and possibly some native restios such as *Baloskion tetraphyllum*, except when the terminal cones are present.

Equisetum arvense.

Environmental effects: the growing of *Equisetum* species has been prohibited in Australia as a precautionary measure, primarily because of their seriousness as potential agricultural weeds and toxicity to livestock. However, common horsetail may also be a potential wetland weed.

Control and management: I can't do better than to quote from my very first book in 1992: these plants 'are indifferent to almost any herbicide invented [and] those few herbicides which will eradicate horsetails are so dangerous it isn't even safe to read their labels'. More recent research suggests that while some widely available herbicides may *sometimes* kill these plants, they may also have no effect whatsoever. It is this pronounced immunity to most herbicides that has created such a furore over their availability through nurseries, though to put this in perspective there have been surprisingly few successes in eradicating other wetland weeds through spraying, either. Most recorded escapes of horsetails were very localised, and have apparently been dealt with effectively in most cases. As their spore needs very specific conditions to grow and is easily killed by drying out, it is possible that horsetails will be completely eradicated within Australia in the near future. The underground runners can travel considerable distances even under concrete or bitumen, so new or previously unrecorded infestations should be reported to the appropriate state authorities, and not tackled by individuals.

Gunnera (Gunneraceae)

The impressively large, spiny-stemmed, broad-leaved herb Chilean rhubarb (*G. tinctoria*; see plate 20) has been recorded as a sporadic garden escape, and although its short-lived seed has fairly specific requirements for successful growth and germination it is regarded as a potential environmental weed by some people. In Australia, this species has been confused with the still larger giant rhubarb (*G. manicata*), which is twice the height at around 3 metres, with broadly scalloped and bluntly lobed leaves compared to the sharply pointed and more deeply divided leaves of Chilean rhubarb.

Origins: cooler parts of South America.

Uses: primarily grown as ornamental garden plants, though the bland stems of Chilean rhubarb are also edible.

Preferred growth conditions: permanently wet but not waterlogged, nutrient-rich soils along stream banks, tolerating some shade. In cultivation both species use up their nutrient supply rapidly and must be heavily fertilised every few years for reasonable growth. Over the past decade of recurring drought in southern Australia both species have disappeared from many gardens, apparently because of intolerance to heat and relatively dry conditions.

Confusing species: the two species can easily be confused while still young, and most plants being sold around the Melbourne area as giant rhubarb several years ago were actually Chilean rhubarb.

Environmental effects: only likely to become minor and localised weeds.

Control and management: even the largest *Gunnera* can be destroyed in a minute or two with just a hatchet, and as all seed produced is likely to germinate within a year or so, any follow-up treatment should not take long.

Gymnocoronis (Asteraceae)

Senegal teaplant (*G. spilanthoides*; see plate 22) is a fairly ordinary-looking, leafy herb that is only easily recognised when it produces its starry, white, ball-shaped flowerheads.

Origins: India, apparently naturalised in central America. In Australia it is now known from many sites in New South Wales and to a lesser degree in Queensland, with other reports from Perth and some as far south as Victoria.

Uses: originally introduced as a rather boring aquarium plant.

Preferred growth conditions: warm, shallow waters and permanently wet soils at the water's edge, rooting freely wherever its stems touch exposed soil, and spreading readily from broken stems. At present, this species is most weedy in disturbed urban situations with high nutrient levels, sometimes growing so densely that it obstructs water flows.

Confusing species: when not in flower, many other introduced aquarium plants, and also some native aquatic herbs.

Environmental effects: not well documented at present, partly because it is not a particularly distinctive plant, and partly because it is only patchily and locally distributed. Senegal teaplant is on the Alert List for Environmental Weeds because there is concern that it could cause comparable problems to the weedier *Ludwigia* species, to be discussed shortly. The most disturbing aspect of Senegal teaplant is that it has already shown it can grow well outside the computer-generated models of its potential range, so we still have no clear idea of just how far it could spread.

Control and management: teaplant is not yet widespread so there is a real hope that this weed could be contained before it spreads to any significant degree. All new outbreaks should be reported to state authorities as it can spread by both stem fragments and seeds. Unlike some other weedy, wetland daisies, this species produces relatively heavy seeds that can float, but are not spread by wind, hopefully confining it to catchments where it is already established. A Weed Management Guide for this species is available online.

Heteranthera (Pontederiaceae)

Mud plantain (*H. reniformis*) is a creeping weed of shallow waters and exposed mudflats, that may rapidly blanket considerable areas with its kidney-shaped leaves; the small, blue flowers resemble those of the related but much larger water hyacinth (*Eichhornia*). *H. zosterifolia* is also grown as a garden plant in Australia, but has not been recorded as a weed.

Origins: northern America.

Uses: sold as an ornamental water garden plant within Australia.

Preferred growth conditions: warm, nutrient-rich soils in shallow water. Although often short-lived in colder climates, it usually grows back from seed in spring.

Confusing species: none.

Environmental effects: regarded as a serious rice field weed in Italy, yet also regarded as endangered within parts of its natural range in the USA. Within Australia it is naturalising in south-eastern Queensland, with some reports from elsewhere, and even in my nursery in southern Victoria it took three years to eliminate this weed from the single pond it was experimentally planted in.

Control and management: the precautionary principle suggests that this minor ornamental should be prohibited throughout Australia as a potentially significant rice field weed.

Hydrocleys (Limnocharitaceae)

Water poppy (*H. nymphoides*; see plate 20) is a strikingly flowered plant with oval floating leaves.

Origins: warmer parts of the Americas, and patchily naturalised in parts of Victoria and New South Wales.

Uses: an ornamental plant grown for its poppy-like flowers.

Preferred growth conditions: warm, shallow waters in full sun, particularly fast-growing in nutrient-rich conditions, but not generally establishing where competing vegetation is already present.

Confusing species: the floating leaves and large, three-petalled flowers superficially resemble those of the native swamp lily (*Ottelia ovalifolia*); however, in the native species the petals are a pure white, very different from the butter-yellow of water poppy. The leaves are also vaguely comparable to those of water hawthorn (*Aponogeton*) which is discussed elsewhere, but the upright, two-columned floral structure of that species is unmistakeable even from a distance.

Environmental effects: a minor and localised weed at the present time, but likely to expand its range if even the most conservative climate change predictions are correct.

Control and management: a precautionary ban on this species should be considered if the incidence of naturalised populations increases.

Hydrocotyle (Apiaceae)

Water pennywort (*H. ranunculoides*) is a round-leaved, creeping plant of shallow waters, with scallop-edged leaves resembling some types of buttercup, and small heads of tiny and inconspicuous flowers that spring directly from the runners. The larger- and rounder-leaved *H. bonariensis* is a common, introduced weed in dunes along the central New South Wales coast (also recorded patchily in Victoria and

The 10-cm diameter leaves of *Hydrocotyle ranunculoides* could be confused with those of some less deeply lobed indigenous species in the eastern states, but this introduced weed is confined to the Perth area at present.

Western Australia), and sometimes also grows in shallow drains, but is unlikely to become a problem in wetlands.

Shield pennywort (*H. verticillata*) is an aquatic native in eastern Australia, prohibited in Western Australia on a precautionary basis, but is not discussed as it is not established as a weed anywhere.

Origins: water pennywort is found from the Americas to Europe, and possibly to western Asia. In Australia, this is a localised though extremely abundant weed in the Perth area, mainly in the Canning River.

Preferred growth conditions: shallow, nutrient-rich soils near the edge of streams and still waters, forming dense mats that may float out from shore.

Confusing species: could be confused with a number of native species of *Hydrocotyle*, especially as the leaves of many of these can vary dramatically in appearance between local forms, but indigenous species in Western Australia have leaves much less than 10 cm wide and are unlikely to be found in water.

Environmental effects: possibly only a serious weed in nutrient-rich streams and disturbed wetlands, as it has shown little sign of moving beyond its present range within Perth. However, the dense growth (which crowds out all other plants along considerable lengths of shoreline) suggests that this is an allelopathic species with potential to create monospecific stands elsewhere if given the chance.

Control and management: the management or control of this weed is a localised problem that does not appear to be spreading rapidly at present.

Hygrophila (Acanthaceae)

Hygrophila species are marsh plants from diverse parts of the tropical world, growing as leafy submerged plants when flooded, and flowering on emergent stems as water levels fall. Of the introduced species, glush weed (*H. costata*) is regarded as the most potentially invasive in Australia, reaching 1 metre and sometimes more in height, and forming densely packed stands comparable to massed alligator weed (*Alternanthera*). However, the variable *H. polysperma* is reported from a similar range and has also been recorded as naturalised further south to around the Sydney area, while *H. triflora* is apparently confined to Berry Springs in the Northern Territory.

Origins: glush weed is originally from South America to Mexico but has apparently been deliberately naturalised in parts of south-eastern Queensland, and also one locality in northern New South Wales, while *H. polysperma* is an Asian plant.

Uses: all of these introduced species were originally imported for planting in aquaria (their flowers are small, packed at the joint of the leaf and the stem).

Preferred growth conditions: warm, shallow waters that may dry out seasonally, sometimes forming monospecific stands on exposed mudflats and other disturbed habitats. Broken stems and even sometimes leaves will take root, and the seed of many species including glush weed is also spread by water.

Confusing species: the emergent growth forms of both native and exotic species can be moderately variable depending on climate, time of year and intensity of light, so it is not always easy to separate introduced and native clones. It should be emphasised that most *Hygrophila* stands found from southern Queensland up to the tropics are far more likely to be indigenous species than deliberately planted weeds. The introduced plants are only known from near urban areas to date, and there is no indication as yet that they are likely to spread over any great distance without human assistance. However, any *Hygrophila* species that look appreciably different from indigenous species already familiar in any particular area should be sent to the state herbarium for identification.

Environmental effects: glush weed has already formed relatively small stands in several Australian locations and has also naturalised in parts of the USA, so it has been suggested that it is an aggressive and invasive plant with real weed potential. *H. polysperma* does not seem to be treated as seriously in terms of weed potential, yet has been recorded further south growing under cooler conditions, suggesting greater adaptability.

Control and management: Given the numerous synonyms and variable growth habits of many *Hygrophila* species under differing conditions, along with the unknown origins and often poorly documented nomenclature of plants sold within the aquarium trade, a blanket moratorium on sale or trade should be imposed for this genus as a whole, until all clones and forms in Australia can be identified and assessed. If suspicions that glush weed has been deliberately naturalised to supply the aquarium trade turn out to be correct, there is every reason to fear that similar attempts will be made with other species, and in this extreme case the only effective deterrent would be to make all *Hygrophila* unsaleable. Manual control of recently established populations is not difficult as they are generally fairly shallow-rooted, though new plants may appear from broken fragments and some species have been reported to grow even from floating leaves. It is not known if any of the introduced species are likely to spread from seed under Australian conditions.

Iris (Iridaceae)

The familiar yellow flag iris (*I. pseudacorus*) with its large yellow flowers has been grown in water gardens for many centuries.

Origins: found from Europe to northern Africa, now patchily established in south-eastern Australia.

Uses: ornamental only.

Preferred growth conditions: shallow waters with a nutrient-rich substrate, though plants will also grow well in moist garden soils and are fairly drought tolerant.

Confusing species: none among aquatic plants.

Environmental effects: a minor weed that spreads by floating seed, sometimes deliberately naturalised.

Control and management: in most cases, even well-established clumps of yellow flag can easily be removed with a backhoe. Free-flowering, complex hybrids between this species and others (e.g. the cultivar 'Roy Davidson') which rarely set seed are widely available, so if self-seeding plants ever become a problem a ban on the true species would impose no hardship on even the most dedicated water gardener.

Lagarosiphon (Hydrocharitaceae)

See the brief discussion under *Elodea*.

Lantana (Verbenaceae)

Lantanas are not wetland plants, but they are among the most invasive, dominating and unpleasant weeds of tropical to subtropical eastern Australia (see plate 24). They are only briefly mentioned here as environmental weeds, which also happen to take over the banks of streams for great distances, crowding out everything indigenous. Although still commonly referred to as *L. camara*, this is a species complex of hybrid origin.

Origins: essentially South America, now a major weed in much of eastern Australia from tropical Queensland to Sydney, and also patchily established elsewhere.

Preferred growth conditions: warm climates with reliable rainfall, growing from full sun to shaded gullies.

Confusing species: none.

Environmental effects: the tropical equivalent of the southern blackberry (*Rubus*), this vile weed adds insult to injury by smelling unpleasant as well. Although lantana thickets aren't avoided by all indigenous animals, they are an impoverished environment at best, and animal diversity wherever lantana has established itself over the past 160 years is undoubtedly much less than it would have been in more pristine conditions.

Control and management: as this is a Weed of National Significance, a detailed Best Practice Manual (Stock 2009) that covers everything from the complex origins of the weeds in Australia, to diverse approaches for control and containment is available both online and in hard copy.

Limnocharis (Limnocharitaceae)

Yellow burrhead or yellow velvetleaf (*L. flava*) is an upright, tropical herb with rounded, fleshy leaves on triangular stems, resembling a yellow-flowered arrowhead (*Sagittaria*) but with the flowerheads bending downwards once they have been fertilised.

Origins: warmer parts of South America to the Caribbean, now locally naturalised in the Cairns to Townsville region of Queensland.

Uses: harvested in Asian rice fields as an edible weed for several specific dishes, and possibly deliberately introduced for this purpose.

Preferred growth conditions: nutrient-rich soils in shallow, warm waters from small pools to the fringes of lakes. A perennial herb to 1 metre high, in ephemeral pools burrhead may grow as an annual, and in rice fields it can become the dominant plant, reducing yields dramatically.

Confusing species: none in its present Australian range.

Environmental effects: mainly known as a significant rice field weed outside Australia, its potential as an invasive plant in relatively undisturbed habitats is not yet known.

Control and management: spreading mainly from seed that may be relatively short-lived, this species has been specifically targeted for eradication before it spreads more widely. Even the smallest previously unrecorded outbreaks should be reported to the relevant authorities immediately.

Ludwigia (Onagraceae)

This genus is a mix of introduced and seriously invasive wetland weeds, water garden and aquarium plants that are not yet environmental weeds but could easily become so, and several native species including one which may be an early introduction to Australia. The proven weeds are marsh ludwigia (*L. palustris*),

Some weedy *Ludwigia* species, from left to right: *L. longifolia* (introduced) with slender, elongated leaves, smooth stems and seedpods; *L. peruviana* (introduced) with markedly furred stems and seedpods; *L. octovalvis* (indigenous) with slightly furred stems and elongated seedpods.

Peruvian water primrose (*L. peruviana*) and the similar-looking long-leaf or willow primrose (*L. longifolia*), with at least two other exotic species still also being sold through the pond and aquarium trades. The ambiguous status of water primrose or clove-strip (*L. peploides* subsp. *montevidensis*) is also discussed here as it has long been treated as a native, but several other species are definitely native as their natural range includes much of south-eastern Asia southwards into the warmer parts of Australia.

Peruvian water primrose (*L. peruviana*) and long-leaved water primrose (*L. longifolia*)

These two species are considered together here as they are superficially similar, and because they are primarily established in the central coastal region of New South Wales. There has also been some confusion caused about these two as the National Alert Weed website lists and illustrates *L. peruviana*, but all of the information on this site clearly and explicitly refers to *L. longifolia*!

Origins: both species originate in South America, and were originally introduced as ornamental plants.

Uses: none. Like many other *Ludwigia* species (as well as other plants in this family), both species are likely to be moderately toxic and are of no value to any known native animals or to livestock.

Preferred growth conditions: shallow, slow-moving or still waters that may dry out in summer, also along streams and in slow-moving waters to around 30 cm deep. *L. peruviana* may form floating mats in still waters, and the seed is probably spread by water; though Sainty and Jacobs (2003) suggest it may also be carried by birds, I have not been unable to find any confirming record of this.

Confusing species and names: the furry-leaved Peruvian water primrose could be confused with the native *L. octovalvis*, also regarded as a minor rice field weed in subtropical and tropical climates, although the range of these species does not overlap in Australia except perhaps to the north of Sydney. *L. peruviana* is usually taller at up to 3 metres, while *L. octovalvis* is generally well under 2 metres. The conspicuous four-petalled, yellow flowers may be up to 3 cm across in both species, but in *L. octovalvis* the fruits are relatively long, narrow and straight-sided to 5 cm or more, while in *L. peruviana* they are much broader and unmistakably furry, reaching around 3 cm.

Long-leaved water primrose (*L. longifolia*) is reasonably closely related to the Peruvian weed, and is also naturalising rapidly around parts of the Sydney area; it has also been reported around Brisbane. Unfortunately, the National Alert Weed website has run descriptions of this species together with Peruvian water primrose, but they are probably comparably weedy. With its elongated, slender leaves on relatively bare, vertical stems, *L. longifolia* is unlikely to be confused with any native *Ludwigia* species anywhere in Australia, so there should be no problem in recognising this weed if it appears in new habitats.

Environmental impacts: in disturbed situations and urban wetlands either of these species could become the dominant plant, producing vast numbers of seeds around autumn in the case of Peruvian water primrose, and shading out most competing growth. In cooler climates they become dormant in winter, leaving nothing but an unattractive forest of dead sticks that is of no value as habitat. These weeds are most vigorous in high-nutrient, disturbed situations and don't establish readily among competing vegetation as the fine seed apparently needs intense light to germinate.

Control and management: already banned, declared noxious or prohibited in many parts of Australia as a precautionary measure against further spread, present-day management programs for both of these species appear to be primarily based on chemical controls. Clearing mature plants will often result in mass germination of seedlings, so this should only be done when the plants are entering dormancy, and an overplanting of dense, low-growing competing and shading plants should be established as quickly as possible after clearing, before seed can start into growth. Deliberate introduction of indigenous pests of native *Ludwigia* species from further north within Australia would be well worth considering, as these would not require the same stringent importation and testing regimes as for biocontrols imported from overseas, and already a native beetle (*Altica* sp.) has been observed feeding upon Peruvian water primrose. In this context, it is worth noting that in Java and Sumatra where this weed was introduced long ago (Soerjani, Kostermans & Tjitrosoepomo 1987), and the same indigenous *Ludwigia* species are present as in northern Australia, Peruvian water primrose is not regarded as a significant weed even in rice fields which suggests natural curbs on its growth and spread are already present.

Swamp ludwigia (*L. palustris*)

This creeping plant is not referred to as a water primrose as its flowers are insignificant, and it is mainly grown for the reddish colouration of the stems and leaves in colder conditions (see plate 28).

Origins: native through much of the northern hemisphere, and in parts of Africa.

Uses: a minor and not very popular ornamental for water gardens, sometimes grown less successfully in aquaria.

Preferred growth conditions: a cool-climate plant that will tolerate hot summers, growing readily on most flooded and waterlogged soils, and spreading readily from broken pieces and possibly seed.

Environmental impacts and values: regardless of origin, many running *Ludwigia* species form dense, floating mats. Anchored at the shoreline these can cover the water's surface for a long distance from shore, so in smaller water bodies they may have a similar smothering effect to the weediest floating plants. Of these, swamp ludwigia is already widely naturalised in cooler parts of the upper Murray catchment in both New South Wales and Victoria, and has the potential to spread much further. Even though it has long been known as a potential environmental weed, there has been no apparent attempt to ban this undesirable plant from sale.

Control and management: the sale or possession of swamp ludwigia and various close relatives (including *L. arcuata* and *L. repens*) should be prohibited throughout Australia as a precautionary move, as it has only naturalised in a relatively small part of its potential range to date. Manual removal is laborious but can be successful in recently established populations. However, feral plants of these species are often submerged or grow in the less accessible parts of wetlands, so spray programs are unlikely to have any significant effect except on the overall health of the wetland.

Water primrose (*L. peploides* ssp. *montevidensis*)

Long regarded as a native, water primrose (see plate 28) was originally described from the Americas, a fair jump for a possibly toxic plant. However, the wide range of many indigenous wetland plants as discussed in Chapter 3 should also be considered in this context.

Origins: native to the Americas, and perhaps also in south-eastern Australia. Like its introduced relatives, water primrose grows readily from cuttings in a nursery situation. It is improbable that any bird could carry a viable piece from the Americas to Australia, and there seem to be no records of birds feeding on this plant, so it is unlikely to be carried in their gut as seed.

Preferred growth conditions: relatively warm and shallow waters, often trailing out over deeper waters, and thriving in more ephemeral wetlands as a bushy but non-flowering herb.

Confusing species: when not in flower water primrose is superficially similar to *L. adscendens*, a more tropical species that ranges from south-east Asia to northern Queensland and the Northern Territory. However, the flowers of the latter species are creamy-white to a very pale yellow with a richer-yellow centre, in contrast to the rich and uniform yellow of *L. peploides*.

Environmental impacts and values: although widespread and sometimes abundant, it is not apparently recorded or known as a food source, or used as habitat by any Australian animals including invertebrates.

Control and management: until its status as an indigenous plant is clarified, water primrose should not be included in wetland plantings (where it can become weedy in any case), and it should not be sold outside its present range.

Mentha (Lamiaceae)

Pennyroyal (*M. pulegium*) is a low-growing, aromatic mint with upright columns of purplish flowerballs on plants to 1 metre tall, spreading rapidly by runners. Other hybrid mints and cultivars (including the sterile hybrid form of *M.* × *piperita* known as Eau-de-Cologne mint) may also be found patchily naturalised in wetlands or even in shallow waters, but are uncommon.

Origins: Europe to western Asia, now widely naturalised elsewhere including in southern Australia, and regarded as a serious weed of agricultural land in Western Australia.

Uses: introduced in the 1800s, possibly as a medicinal herb.

Preferred growth conditions: moist to seasonally wet, often sandy soils but does not thrive in waterlogged conditions.

Confusing species: when not in flower can be confused with some other species of *Mentha*, possibly including several smaller-leaved indigenous species that are found on the fringes of wetlands and streams, but rarely grow as vigorously or as tall.

Environmental effects: primarily a weed of disturbed ground including poorly drained pastures.

Control and management: pennyroyal is unlikely to be found in wetlands, only on damp ground in their vicinity, and is most weedy in wet, poorly managed pastures. Control should focus on preventing plants from flowering and setting seed, although root fragments can also be spread in the course of ploughing.

Mimosa (Mimosaceae)

Mimosa or giant sensitive plant (*M. pigra*; see plate 26) is a fast-growing, prickly stemmed shrub to 6 metres tall, with paired ranks of leaflets resembling some native wattles (*Acacia* species). The rounded, 2-cm diameter clusters of pink to mauve flowers mature to form hairy pods that fall apart as they ripen, leaving just a curious external frame resembling a safety-pin.

Origins: tropical America, now a major weed in parts of Africa and Asia, as well as in the Northern Territory where it forms monospecific stands over many thousands of hectares, with a localised outbreak also recorded in northern Queensland.

Uses: apparently first grown as an ornamental in the Darwin Botanical Gardens in the 19th century, or perhaps introduced as a novelty, as the leaves close together when touched.

Preferred growth conditions: although this weed tolerates seasonal flooding, it is mostly found on moist to wet ground on the higher levels of floodplains and along streams.

Confusing species: could be confused with some other woody plants of this family when not in flower, particularly some wattles (*Acacia*), though these are not generally found in comparable habitats and are not as notably spiny.

Environmental effects: mimosa forms dense, prickly stands that can shade out virtually all other plants, and will even establish in paperbark (*Melaleuca*) swamps where it grows so densely that it can prevent recruitment of paperbark seedlings.

The resulting greatly simplified ecosystems support only a fraction of the animal diversity found in natural wetlands.

Control and management: a serious though still relatively localised weed given its considerable potential range in tropical Australia, and difficult to control once established. New infestations should be reported to authorities for appropriate action, though small infestations can be successfully removed by hand if caught early enough, ideally before the plants have had a chance to set seed. Mimosa mostly spreads by floating segments of the seed pods, each containing a single seed, though broken stems will also take root. Large plants can produce hundreds of thousands of seeds each year, and these may remain viable for 20 years and more. Water buffalo and feral pigs have been significant agents for the spread of this species in the past, creating new disturbed habitat ideal for seedling germination, though this problem has been abated to some degree by shooting programs in more recent years.

Control by cutting and painting with appropriate herbicides works well to control established trees, but new infestations from seed are likely to follow. A number of insects have been tested as biological control agents, though the results have not been dramatic to date. Moreover, due to the disappearance of indigenous vegetation from many sites where mimosa has been successfully controlled in recent years, other weedy species (particularly the grass *Urochloa*) are often the first plants to reestablish. A Mimosa Weed Management Guide is available on the Weeds of National Significance website.

Myriophyllum (Haloragaceae)

Parrotfeather (*M. aquaticum*; see plates 19 and 22) is a feathery-leaved, blue-green creeping plant of shallow waters, often forming a plush mat of foliage above the water's surface.

Origins: South America, but now widely entrenched in eastern Australia and patchily in the west. This is a noxious species in Western Australia, and prohibited in Tasmania.

Uses: introduced as an ornamental plant.

Preferred growth conditions: nutrient-rich, shallow waters, tolerating fairly shaded situations and also full sunlight.

Confusing species: some native water milfoils (*Myriophyllum*), though these can be separated by a close comparison of the foliage alone.

Environmental effects: forming a dense-growing emergent carpet, parrotfeather is possibly ecologically equivalent to some of its larger native relatives, providing shelter from predators for smaller fishes and tadpoles. However, in many urban situations it grows so densely that few animals (even including invertebrates) will live or breed in the resulting thickets.

Control and management: although this weedy species can be extremely vigorous in ideal conditions, it is not difficult to remove most plants by hand in smaller areas. A further check a few months later should be enough to get rid of any surviving stems, and as only female plants (probably just a single clone) are present in Australia it will not return from seed. Given that this clone only reproduces by stem fragments, it is possible that with time and increasing accumulation of viral and other disease organisms, it will eventually disappear spontaneously.

Other exotic *Myriophyllum* species: In the mid-1990s I was sent plants of Eurasian watermilfoil (*M. spicatum*) from a water garden nursery in Sydney. This species had been imported in the 1960s, at a time when there was little restriction on the aquatic and wetland species that could be brought in, and the nursery in question destroyed these plants at my suggestion. However, other nurseries may still inadvertently be selling this potentially serious weed, as some offer dubiously identified watermilfoils including 'natives', for which neither the photos nor the descriptions match up with the names used.

Nasturtium (Brassicaceae)

Watercress (*N. officinale*; see plate 27) is a fast-growing herb with deeply divided, paired leaves on upright stems, topped with small heads of white flowers. The very similar-looking one-rowed watercress (*N. microphyllum*) is apparently much less common, though it may just go unnoticed when growing with common watercress, and is restricted to the cooler parts of south-eastern Australia.

Origins: both species are Eurasian in origin, and watercress has been long naturalised across most of Australia in cooler, flowing waters.

Uses: deliberately introduced as an edible plant for its high-quality, slightly peppery leaves and stems. However, these should only be harvested from clean, flowing waters.

Preferred growth conditions: brightly lit situations in cool, relatively shallow, slow-moving streams and open wetlands, often in slightly alkaline waters.

Confusing species: related cresses including *Rorippa*, though most of these are considerably smaller and grow on higher and drier ground, rarely in water.

Environmental effects: although an abundant plant in many situations, watercress is apparently fairly neutral in its environmental impacts and may even be eaten by some waterbirds in small amounts. It is also used as shelter by a range of fishes, and sometimes by tadpoles of stream-dwelling frogs.

Control and management: there is no prospect of eliminating watercress, as it has probably colonised every potential habitat available to it from the fringes of deeper pools and lakes, to eutrophic streams flowing through cattle country, and stunted plants can even be found in seeps just above the high tide mark along southern coasts. Mature plants may sprawl over many square metres, producing literally millions of seeds, and some of these are likely to still be viable a decade and more later (the family as a whole is noted for its long-lived seed).

Nuphar (Nymphaeaceae)

Spatterdock or brandybottle (*N. lutea*; see plate 19) is a close relative of the more ornamental waterlilies, with similar though more elongated leaves and relatively small, globular flowers with a curious odour likened to fermenting fruit.

Origins: cooler parts of the northern hemisphere.

Uses: introduced as a water garden curiosity.

Preferred growth conditions: cool, often deep waters to 3 metres and more, on nutrient-rich soils.

Confusing species: when not in flower, the true waterlilies (*Nymphaea*).

Environmental effects: probably comparable to *Nymphaea*, to be discussed shortly, but capable of growing as vigorously as the weedy Mexican waterlily.

Control and management: spatterdock was the dominant submerged plant in lagoons of the Melbourne Botanical Gardens two decades ago, ultimately covering several hectares, and it isn't clear how much it ultimately cost to eradicate it. It is not yet naturalised in Australia, but given the extremely low level of interest in this and related species, and the difficulty of controlling any of them in deeper waters, a pre-emptive ban on all *Nuphar* species is unlikely to arouse much opposition.

Nymphaea (Nymphaeaceae)

Waterlilies are among the best known aquatic plants, so widely naturalised (see plates 1 and 18) that few people realise that most of the varieties in dams and urban streams across much of this country – and even some of those in wetlands – don't actually belong there. The handful of indigenous waterlily species are relative newcomers, descended from south-east Asian species that have adapted to northern Australian conditions over the past few million years. If you live well south of the Queensland–New South Wales border you can be sure all waterlilies in your area are introduced, but in the Northern Territory and less reliably in tropical Queensland, native species are often abundant, and identification of any waterlilies present is essential before you even think in terms of weeds.

The following discussion is entirely concerned with introduced waterlilies, as these are often deliberately naturalised in still or slow-moving waters, thriving on nutrient-rich soils, even where livestock have free access. Any move to eradicate introduced waterlilies from urban landscapes and farm dams would almost certainly arouse significant opposition, and not just from gardeners. This is not necessarily a problem, as unlike willows (*Salix*), which are less than worthless as riparian habitat, the effects of many waterlilies are relatively innocuous.

Instead, control and regulation of waterlilies should focus on those particular species, groups and hybrids that could pose a real or potential threat to the ecology of relatively undisturbed wetlands. Waterlilies are considered here as two major groups, with very different cold tolerances, growth patterns and weed potential, known to gardeners as 'tropicals' and 'hardies' though their potential growing ranges overlap considerably in Australia. Due to the maze of related issues which needs to be negotiated, these distinctive groups are discussed in essay form, rather than in the simpler format used for nearly all other plants in this book.

Hardy waterlilies

Hardy waterlilies as defined by water gardeners are all species and hybrids descended from temperate zone, northern hemisphere species, some of which range as far north as Scandinavia. Not surprisingly, all of these thrive even in very cold climates, with the foliage dying away during the colder months, and they are largely unaffected by sub-zero temperatures. Floras usually treat these plants as just variants of several species, but naturalised hybrids are not uncommon and are also more diverse, varying in colour from white to yellow, pink and even crimson.

The flowers of the various cultivars vary considerably and although in most hardy types these rest on the surface of the water, in some cases they may be held above the surface like a tropical waterlily. The rhizome growth of the species used to breed new cultivars also varies considerably, from relatively long and straight, to twisting and branching repeatedly. In some complex hybrids, the rhizomes will ultimately grow into a bizarre gordian knot unless the plant is divided and replanted every few years.

Mexican waterlily (*N. mexicana*; see plates 1 and 18) is the weediest of the hardy species, and sometimes spreads to cover many hectares in waters up to 3 metres deep. In ecological terms it is intermediate between the hardies and the tropicals but does not hybridise with the latter group, and retains its leaves long after most hardies have become fully dormant. Most yellow- or cream-coloured hardy hybrids include this species in their ancestry, though the hybrids are usually far better mannered. Mexican waterlily is still sometimes offered as an unnamed yellow waterlily, a good reason in itself for prohibiting the practice of selling waterlilies by colour alone.

With at least 120 named hardy cultivars already available in Australia, most of them non-weedy, it is perhaps time to reconsider the importation of further hardy waterlily imports into this country. The USA has been the major centre for the breeding of new hardy waterlily hybrids for the past few decades, and breeders there have put much effort into importing new, vigorous strains and wild species from many parts of the world. These are much more fertile than the ageing clones that were used to breed the 'classical' hybrids as much as a century ago, and some of them can set considerable quantities of viable seed. If naturalised, such plants could potentially create a new type of weed that could spread to new places in southern Australia, without being deliberately planted.

Hardy waterlilies probably have some value as habitat, though few if any southern Australian animals have evolved with them. Southern fishes (mostly evolved from marine ancestors) may use waterlilies as shelter from predators, but there are many indigenous plants which are at least as useful for this purpose, and while southern frogs descended from tropical species may recognise waterlilies in some dim, archetypal fashion they don't seem to seek them out. Non-weedy hardy waterlilies are acceptable in dams and in disturbed urban situations, where they help keep water temperatures down in summer, but we don't need new varieties capable of spreading into relatively undisturbed wetlands from seed.

Tropical waterlilies

Unlike most hardies, tropical waterlilies hold their flowers well above the water on upright stems and their colour range is different, including blue, purple and even greenish types. Some introduced tropical waterlilies are closely related to the native species of northern Australia, so they provide a type of habitat which many native fishes, frogs and many birds are already familiar with, but they almost certainly also compete for similar niches in natural wetlands. Their fruits are recognised as food by herbivorous waterbirds that already feed from indigenous waterlilies, which means that their seed is probably being intermittently delivered into natural wetlands in waterbird droppings, though if this is the case it may be some time before it becomes obvious.

Few people are aware of differences between most blue or white tropical waterlilies, so introduced tropical species could easily become a cryptic plague in northern Australia, much as *Typha latifolia* has already become in the south. It is not just the introduced species that have the potential to spread from seed, as many of their hybrids are at least as fertile. The sale of tropical waterlilies has already been restricted in parts of northern Australia, but most of the pretty blue waterlilies on the highway between Brisbane and Gladstone in central Queensland are already introductions – most commonly Cape waterlily (*N. caerulea*; see plate 18) from southern Africa, with outlying feral populations already found as far south as Sydney.

Pistia (Araceae)

Water lettuce (*P. stratiotes*; see plate 16) is a fast-growing, floating plant with velvety leaves arranged in a tightly packed rosette, multiplying by runners that spread along the water's surface, and sometimes from floating seeds which may drift before settling and germinating.

Origins: a pantropical species probably native in the Northern Territory where it is nowhere near as weedy, suggesting that it is already integrated into local ecosystems, but introduced in warmer areas of eastern Australia.

Uses: a minor ornamental, banned as a potentially noxious species in New South Wales. Although eaten to a limited degree by livestock, it is low in protein and is unlikely to be a useful stockfeed.

Preferred growth conditions: relatively still, warm waters with abundant sunlight, though in tropical climates water lettuce may still grow vigorously even if heavily shaded for much of the day. Plants will also take root in mud if water levels drop, but don't spread rapidly under such conditions. Although reputed to be frost-sensitive and unable to survive much south of Brisbane, the seed will remain viable for several months in surprisingly cold water, and survives freezing for several weeks. An over-wintering trial I carried out on mature plants in a small, outdoor pond at 250 metres altitude in southern Victoria suggests some degree of cold tolerance, as most plants were still alive at the end of winter, though a late frost killed nearly all of these a few weeks later.

Environmental effects: in warmer climates and still, nutrient-rich waters water lettuce can rapidly form a smothering blanket over large areas, potentially creating mosquito habitat and reducing dissolved oxygen levels dramatically. If the mature plants are killed off by cold weather, the ever-increasing mass of dead vegetation reduces dissolved oxygen levels still further as it decays.

Control and management: the relatively restricted natural range in the Northern Territory indicates that this species does not spread readily between wetlands, and all occurrences elsewhere in Australia are likely to be deliberate introductions by humans. However, projected global warming scenarios suggest that this species will be able to grow much further south in the future, and a wider precautionary ban on its sale should be considered.

Pontederia (Pontederiaceae)

Pickerel rush (*P. cordata*) resembles some forms of *Sagittaria* with heart- to lance-shaped leaves on upright stems, but the flowers of most varieties in cultivation form a blue or lilac bottlebrush held above the foliage.

Origins: the Americas, naturalised sporadically in both eastern and western Australia.

Uses: primarily an ornamental plant.

Preferred growth conditions: nutrient-rich soils in relatively permanent waters up to 0.5 metre deep, often dying out during dry periods.

Confusing species: the foliage is superficially similar to *Alisma* and some *Sagittaria*.

Environmental effects: usually only naturalised in urban situations, with at least some of the feral populations disappearing in southern Australia during the recent period of prolonged drought.

Control and management: the most familiar blue strain of pickerel rush does not appear to set viable seed, and has shown little sign of becoming weedy under Australian conditions. However, many new forms with flower colour varying from blue to pink and even white have been introduced from seed over the past two decades, and if cross-pollination makes spread by seed possible the weed potential of this species may need to be reassessed.

Populus (Salicaceae)

White poplar (*P. alba*; see plate 25) is a moderately large tree widely grown in southern Australia, though in more recent decades the very upright Lombardy poplar (*P. nigra* 'Italica') has replaced it completely as a windbreak and avenue tree.

Origins: both species in their typical forms are found from Europe into Asia and northern Africa.

Preferred growth conditions: wet soils along streams and around ponds or dams, where white poplar may grow prolifically in a relatively stunted form, tolerating occasional flooding.

Confusing species: the silver-backed, oval to slightly divided leaves and distinctively patterned bark of white poplar are very different from all other deciduous trees found in riparian situations.

Environmental effects: where poplars grow densely they will often displace native plants, though not to the same extent as willows (*Salix*), and there is little evidence that they are likely to invade relatively undisturbed wetlands. However, they are just as thirsty as willows and dense stands increase evapo-transpiration, reducing environmental flows.

Control and management: even mature trees are relatively easy to kill off by the same methods as willows if necessary.

Ranunculus (Ranunculaceae)

Celery-leaved buttercup (*R. sceleratus*) is the only introduced wetland species of this group to have naturalised on any significant scale, a tiny-flowered, short-lived herb to less than 1 metre in height, with fairly narrow, deeply divided leaves. Other introduced species (not discussed further) include creeping buttercup (*R. repens*), which may form dense carpets over moist soils but is rarely found in wetland situations except in drains and other disturbed situations. The water garden plant *R. flammula* has been reported as naturalised in New South Wales, but only on a small scale to date. Given that the total sales of the latter plant Australia-wide are on the order of several hundred dollars annually, and that it is also toxic to livestock, a pre-emptive ban on this visually unexciting species would not be likely to arouse any interest from gardeners.

Origins: the Eurasian celery-leaved buttercup is now widespread in south-eastern Australia, and is reputed to be the most toxic of this group, which includes many other unwholesome species.

Preferred growth conditions: an annual species growing on disturbed ground that may be flooded at times, including pastureland, and only occasionally found in relatively undisturbed natural wetlands.

Ranunculus sceleratus.

Confusing species: small or stunted, non-flowering forms of some other buttercups, including indigenous species.

Environmental effects: primarily a weed of disturbed pastures, drains, and the fringes of created wetlands, growing most vigorously in high-nutrient situations, and mainly of concern when it becomes a threat to livestock.

Control and management: the seed seems to be fairly short-lived, so manual removal of plants before they have flowered can dramatically reduce infestations, and follow-up treatments over the next two years should eliminate it.

Rotala (Lythraceae)

R. rotundifolia (see plate 20) is a creeping herb of shallow waters with round leaves and short columns of pink flowers, possibly toxic as the leaves are left untouched by a wide range of animals from aquatic snails to herbivorous waterbirds.

Origins: found from India to Japan, and recorded as naturalised on a small scale around Brisbane and Sydney.

Uses: an ornamental only.

Preferred growth conditions: shallow, warm waters, but also tolerant of light frosts, growing vigorously even at fairly high altitudes in southern Victoria.

Confusing species: none.

Environmental effects: not known, but if this species appears from seed beyond the urban areas where it is recorded at the present time, it should be treated more seriously as a potential weed.

Control and management: difficult to remove once naturalised as the slender, running stems are usually interwoven with other vegetation.

Rubus (Rosaceae)

Blackberries are a complex of hybrids and species that are often the dominant vegetation along the disturbed shores of southern Australian streams, comparable in effects in their southern range to lantana (*Lantana*) in more northern areas. Although blackberries aren't wetland plants they are briefly discussed here in the context of their impacts on riparian communities (see plate 24).

Origins: from temperate Europe, Asia and North America, a diverse and complex range of plants now widely naturalised even in the southern hemisphere.

Uses: introduced primarily because of their edible fruits, which vary widely in quality according to species, hybrid, and even growing conditions.

Preferred growth conditions: lightly shaded soils in areas where humidity remains reasonably high throughout the year, and not just near wetlands or streams.

Confusing species: there are also several native *Rubus* species, so accurate identification is essential before control measures are considered.

Environmental effects: as for lantana, blackberries smother most indigenous plants, make access to water difficult for any animal larger than a rabbit, and completely alter the ecology and appearance of thousands of kilometres of riparian habitat.

Control and management: see the definitive guide on blackberries available as a download or in hard copy from the Weeds of National Significance website (see Anon 2009) for a range of control measures and potential management strategies. Localised infestations are not difficult to clear roughly with a slasher or brushcutter, but cut stems should not be left in contact with soil or they will take root again. Young regrowth is relatively easily to pull out repeatedly with a gloved hand or a pair of pliers, but this process may need to be repeated over many months before the underground reserves are permanently depleted. On a larger scale, my preferred long-term strategy for reduction of introduced blackberries is overplanting with locally appropriate shading plants such as blackwood (*Acacia melanoxylon*), which eventually reduce light levels below what blackberries can survive on, allowing regeneration or replanting of some of the shade-tolerant species such as ferns which formerly grew in these low-light environments.

Rumex (Polygonaceae)

Curled dock (*R. crispus*) and several other of its introduced relatives are primarily weeds of pasture, agricultural and wastelands, and are only likely to be found around the fringes of natural wetlands. Several of the introduced species are declared noxious weeds in some parts of Australia, and the native swamp dock (*R. brownii*) is also a declared pest plant in Western Australia when growing on agricultural land, though is possibly a minor habitat plant in natural wetlands.

Origins: the various introduced docks are northern hemisphere plants, mainly from Europe and western Asia.

Preferred growth conditions: nutrient-rich, moist to wet soils in disturbed places, thriving particularly well where competing vegetation is kept down by grazing.

Confusing species: various indigenous docks also grow on the fringes of wetlands, and the native mud dock (*R. bidens*) even in permanent waters, so accurate identification to species level is essential before any action is taken against apparent weeds of this genus.

Environmental effects: minor weeds in relatively undisturbed wetlands.

Control and management: usually only forming a narrow fringe around wetlands in readily accessible places, where they can readily be dug out if only a small number of plants is involved, including as much of their thick roots as possible. Seed will continue to germinate over the next few years, so follow-up treatments should be done annually before these mature, in spring or autumn when most seed germinates.

Sagittaria (Alismataceae)

Arrowheads are close relatives of water plantains (*Alisma*), named for the projecting lobes of many species which resemble a large arrowhead; however, in other species the simpler leaves are closer to water plantain in appearance. Giant

Sagittaria platyphylla.

arrowhead (*S. montevidensis*; see plate 30) is fairly typical of the true arrowheads, while the leaves of delta arrowhead (*S. platyphylla*) are simple spearheads. Several *Sagittaria* species still being sold through the water garden trade could also become minor weeds in inland irrigation channels and possibly elsewhere, but have not been recorded as naturalised to date.

Origins: arrowhead was introduced from South America and is now naturalised in south-central New South Wales and in north-central Victoria. Delta arrowhead is a Mexican and North American species more widely spread as far south as central Victoria, with isolated records from Western Australia and around the Sydney to Newcastle area.

Uses: tubers of all *Sagittaria* species are edible after cooking, though slightly bitter.

Preferred growth conditions: shallow to moderately deep waters to 50 cm deep, growing most vigorously in areas where regular seasonal fluctuations retard the growth of competing plants that can't die down to tubers.

Confusing species: immature leaves of the various arrowheads are similar to those of water plantains but their stems are much thicker, softer and fleshier in proportion to leaf size, and both naturalised species produce much larger, three-petalled white to pinkish flowers in relatively small numbers.

Environmental effects: despite some claims that arrowheads compete actively with indigenous plants, these are almost entirely weeds of disturbed situations such as open irrigation channels, although isolated patches may also sporadically appear in backwaters of streams and rivers. Arrowheads may also appear in rice fields but are not usually persistent, and according to some sources don't markedly affect yields except if they have been allowed to grow out of hand through poor management. Both species are prohibited in many states for obvious precautionary reasons, but are unlikely to become significant weeds in undisturbed natural wetlands.

Control and management: the seed is short-lived so arrowheads may be among the most ephemeral of aquatic weeds, disappearing from regularly cultivated rice fields, and only reappearing as seedlings from plants already established in the supply channels. These are more persistent as they can grow back from dormant tubers after dry spells, though some sources suggest that two consecutive years of drought may be enough to even kill these.

Salix (Salicaceae)

Willows have been a part of wetland landscapes in southern Australia for so long that most people aren't consciously aware of them as an invasive blight, and in many cases have a liking for them. As with lantana (*Lantana*) and blackberries (*Rubus*), willows are so familiar to people living where such weeds are abundant that they don't require any detailed description (see plate 25). Their control and management is only briefly outlined here, as there is already a considerable and readily available literature.

Origins: temperate regions of the northern hemisphere; many naturalised willows in Australia are of hybrid origin.

Uses: introduced to lend an English feel to the landscape, and widely planted to hold eroding streambanks together after the original, indigenous vegetation was removed. However, as soft-timbered hardwoods, willows do a much worse job of erosion control because they are relatively brittle and will often break during serious flood events, while larger trunks can even act as a lever that collapses steeper banks as the trees twist and fall.

Preferred growth conditions: most willows are basically riparian plants, although they will also grow in any permanently moist soils well away from streams and wetlands.

Confusing species: there are many species and forms of willows still being sold legally, from larger shrubs to trees, as the various states have yet to agree on a complete abolition of all willows because some are still regarded as ornamental plants. Yet their economic value in terms of nursery sales is unimpressive compared to the ever-escalating cost of control programs, and even such apparently innocuous species as the largely self-sterile weeping willow (*S. babylonica*) can produce considerable amounts of hybrid seed when growing near other species that can pollinate it.

Environmental effects: willows are worthless as riparian habitat for virtually all indigenous animals, dump masses of semi-toxic leaves into rivers at the end of their growing season, contribute to the collapse of river banks during periods of flood, and transpire immense volumes of fresh water during hot weather, significantly reducing environmental flows. They radically change riparian habitats, shading out nearly all of the indigenous plants that once grew there, and create new types of erosion problems as they are adapted to more permanent streams rather than the often variable flows of Australian rivers.

Control and management: control of willows should be done in sections, ideally working downstream so that new infestations don't replace the removed trees from snapped-off twigs or branches, or from seed. The simplest method is cutting established trees and painting the wound as quickly as possible with an appropriate herbicide, but this must also be combined with replanting with appropriate indigenous plants. If other invasive weeds are also present in the same area removal of willows may aid their spread (see for example comments under *Glyceria*), and this should be allowed for in any willow removal/control plan.

Various biocontrols are under consideration, and the willow sawfly has spontaneously appeared (apparently wind-blown from New Zealand) around 2004 in eastern Australia; it is now also recorded from Western Australia. This willow-eating insect can cause significant damage to larger willows following warm, dry spring conditions. For more detailed information on the many willow control options and strategies available, see the Willows Management Guide (Holland-

Clift and Davies 2007) available from the Weeds of National Significance website online, or in hard copy.

Salvinia (Salviniaceae)

Salvinia (*S. molesta*; see plate 17) is a distinctive floating fern with rounded leaves that grows so vigorously it has carpeted entire lakes and reservoirs in just a few years.

Origins: Brazil, now widely naturalised and regarded as a serious weed in many tropical countries.

Uses: introduced as an ornamental plant, salvinia is now a prohibited weed throughout Australia, and possibly naturalised in all mainland states though only on a limited scale in southern Australia at the present time.

Preferred growth conditions: warm, relatively still waters and tolerant of moderate salt levels, in southern Australia sometimes surviving light frosts.

Confusing species: although salvinia varies in appearance to some degree depending on growing conditions, it can easily be recognised from the rounded leaves cupped around soft, waxy hairs (though the leaves can be more crowded and compressed in densely growing mats), and is unlikely to be confused with any other floating plant in Australia.

Environmental effects: like many floating plants this fern can grow extremely rapidly in high-nutrient situations, forming a dense mat that cuts off light and oxygen to the waters below, and creating depauperate communities of just those few animals which can either breathe air directly or can survive with very little dissolved oxygen. Because of its impressive rate of spread this is one of the most undesirable weeds in water supply catchments, not only reducing water quality under the blanketing surface layer, but also contaminating it as ageing plants die, sink and decay.

Control and management: this species of salvinia is believed to be infertile so it has almost entirely been spread by deliberate introduction, and possibly sometimes by waterbirds over short distances. Manual removal is unlikely to be successful except in smaller water bodies, as even the smallest remaining piece can regrow rapidly. Biological control by weevils introduced to deal with this species has worked spectacularly well on larger infestations in warmer parts of Australia, destroying the species completely on some lakes. As this weed is probably a single clone, in the long term it will hopefully die out (or at least be greatly reduced in vigour) as it accumulates viral and other disease problems.

Shinnersia (Asteraceae)

Mexican oakleaf (*S. rivularis*) is a marsh herb to around 1 metre in height, with variably divided leaves in the emergent form, and small white flowerheads.

Origins: Mexico to the southern USA, in recent years being increasingly promoted as a cold-water aquarium plant in both Europe and the USA.

Uses: two distinct forms have been introduced recently (and possibly illegally) into the water garden trade within Australia.

Preferred growth conditions: shallow waters and the fringes of wetlands, surviving frosts and reported to grow vigorously even outdoors in England during the summer months.

Confusing species: when not growing in its emergent form or in flower, the relatively simple, rounded submerged leaves resemble those of many other species of aquarium plant.

Environmental effects: not yet recorded as naturalised in Australia, but alarmingly already a naturalised weed in several central European countries. On aquarium website forums it is also already being described as one of the fastest-growing aquarium plants known, even under the extreme low-light conditions of the artificially lit aquarium! A recent proposal to ban this species as a potential weed in Texas was overturned when it was realised that it is native there.

Control and management: Mexican oakleaf should be prohibited before the inevitable dumped plants make their way into the wider environment.

Veronica (Scrophulariaceae)

Blue water-speedwell (*V. anagallis-aquatica*; see plate 27) and pink water-speedwell (*V. catenata*) are upright herbs of seasonally flooded places, growing to around 1 metre high, though plants may be taller in nutrient-rich soils.

Origins: Eurasia, now also patchily naturalised through south-eastern Australia; blue water-speedwell is the more widespread of the two species.

Uses: the leaves of blue water-speedwell are edible raw or cooked.

Preferred growth conditions: permanently moist to wet soils which flood at times, sometimes deeper in clear, slow-moving waters. Neither species seems particularly tolerant of heat, and in their more northern range within south-eastern Australia they are more likely to be found at slightly higher and cooler altitudes.

Confusing species: can be confused with some naturalised aquarium plants when not in flower.

Environmental effects: these are relatively minor and localised weeds, growing most vigorously in disturbed situations, although they may also sometimes appear on the fringes of wetlands.

Control and management: manual removal of recently established plants will usually prevent further spread from seed.

Wachendorfia (Haemodoraceae)

Bloodroot (*W. thyrsiflora*; see plate 21) is a curious plant with deeply pleated, palm-like leaves, slender columns of small, yellow flowers reaching 2 metres in height, and dense mats of red, fibrous roots.

Origins: southern Africa.

Uses: widely grown as an ornamental.

Preferred growth conditions: permanently moist to waterlogged soils, where it sets copious quantities of large seeds which germinate readily.

Confusing species: the combination of pleated, palm-like leaves and tall pokers of yellow flowers is distinctive.

Environmental effects: probably comparable to *Aristea*, which has already been discussed.

Control and management: see comments for *Aristea*.

Zantedeschia (Araceae)

Arum lily, calla lily or funeral lily (*Z. aethiopica*; see plates 1 and 21) is a broad-leaved herb growing to around 1.5 metres in height, although smaller cultivars not much over 50 cm are also sold in Australia. A cultivar called 'Green Goddess' has a variable green flower border, but is otherwise identical to the typical forms and comparably invasive. Smaller forms with disproportionately large flowers such as 'Childsiana' and 'Crowborough' may be less weedy.

Origin: introduced from southern Africa as an ornamental.

Uses: an ornamental, though not greatly in demand due to its association in many cultures with funerals. Although it has been reported as a famine food, all parts of this plant are toxic to some degree, and it is neither used nor inhabited by any native animals.

Preferred growth conditions: grows most vigorously in any wet or permanently moist soils, particularly in warmer and subtropical areas. Arum lily will tolerate flooding for long periods of time, but is also very drought tolerant and can survive for many years as dormant rhizomes in drier soils. The seed is easily dispersed downstream while fresh, germinating on moist soils where it has been carried by water or has fallen during the cooler months.

Confusing species: the wide, white 'flowers' with a yellow spike in the centre are unmistakable, and are unlikely to be confused with any other introduced or native flower found in wetland situations.

Environmental effects: a common and widespread weed in southern Australia, still being planted as it is not recognised as a serious environmental problem outside Western Australia, where it forms single-species stands along long sections of some brooks and wetlands. In cooler climates in the eastern states it is not as abundant, but several degrees of global warming would most certainly change this. Arum lily should be declared a noxious and unwelcome environmental weed throughout Australia, as most of the relatively small and localised outbreaks in the eastern states are still potentially controllable at the present time.

Control and management: seed can be killed by drying out and its viability is short, so slashing or brushcutting at the beginning of the flowering season may be a workable short-term management strategy. As these plants are often found well within wetlands (particularly in Western Australia), spray programs of any kind will have impacts on overall wetland health, and at least some of the tubers are likely to become dormant in response to herbicides so all sprays which have been recommended for control of this species would need to be repeated over many years – with unknown effects on the wider environment.

Algae and cyanobacteria

Ideally, this book should have included a discussion of some of the more significant genera of freshwater algae, as well as some of the better-documented exotic seaweeds, but in the course of writing it became clear that any detailed discussion of these very different plants could not be easily fitted into a book concerned primarily with wetland weeds. One complicating factor was that so many groups of inland algae are cosmopolitan that there is no way to decide whether they are also indigenous, and as their role as weeds (or otherwise) must hinge on this crucial aspect it was decided to reduce all discussion of the various groups of algae to a brief overview.

Many microscopic algae along with the still-smaller cyanobacteria ('blue-green algae'; see plates 30 and 31) combine to form floating mats in disturbed situations, and it is these conglomerates which usually attract public attention rather than their less conspicuous roles in the natural world. The vigorous growth of such species is usually a response to intense light combined with an unnaturally high level of dissolved nutrients, often where there is fertiliser runoff from agricultural land. The simplest (though unreliable and often slow) cure is overplanting with other nutrient-absorbing plants that take advantage of the available nutrients, while also reducing light levels below.

Even in more natural situations where small-celled algae often go unnoticed, they may still be valuable sources of food for a diverse array of similarly small and equally poorly studied invertebrates. The combination of extremely fast algal reproduction (particularly of planktonic species) with comparably vigorous predation means that the standing crop in an undisturbed wetland may not necessarily be impressive at any particular moment in time, but can sustain a surprisingly large numbers of planktonic animals. In turn, these are among the most important foodstuffs for a wide range of larger animals.

Some of the more conspicuous types of anchored freshwater algae (*Chara* and *Nitella*) have already been discussed along with other indigenous plants, due to their apparently complex structure, but the smaller species are more likely to form floating mats which may be a mix of species. Apart from the unsightly nature and frequently malevolent colours of algal mats, it is their possible toxicity that disturbs most humans. Some of the most toxic allelochemicals are produced by algae and even more so by cyanobacteria, and there is reason to think that most types of algae not only use chemical weapons against each other, but also against any other organism that they wish to deter from feeding upon them.

Algal effects on other plants may sometimes be incidental to their wars with each other and with their predators, but in a newly constructed wetland with high nutrient levels, a rapid proliferation of algae may reduce the range of wetland plants that can survive. On the more positive side, cyanobacteria fix nitrogen direct from the air and make this available to 'higher' plants in a useful form, while also binding and possibly to some degree neutralising some of the raw and least wholesome ingredients of recently flooded terrestrial soils.

Within Australia, the best and most practical identification guide for freshwater and inland algal species is Entwisle, Sonneman and Lewis (1997). Although you will need strong magnification to see the relevant details for most species, inexpensive USB microscopes that project directly to a computer screen for photography are becoming increasingly available at the time of writing.

Seaweeds

Seaweeds include a number of (sometimes distantly) related marine algae, and most of these are more complex than the majority of freshwater algal species. It is likely that dozens of exotic seaweeds have naturalised in Australian waters to date, though some are so widely distributed that the Australian populations could have come from almost anywhere, including pre-1788 introductions from the hulls of wooden boats. Lewis (1999) lists more than 20 'probably' introduced species for southern Australia, along with another 50 cosmopolitan species recorded for Port Phillip Bay alone, which are only included as 'possible' introductions. Only two seaweeds are discussed briefly here, selected to show the complexities of controlling weeds that arrive by sea, and even deciding if they should be treated as weeds.

There is no doubt that many seaweeds have been introduced either in bilge water released from ships, or arrived attached to their hulls. Of these, Japanese kelp (*Undaria pinnatifida*, an edible species sold as wakame in Japan; see plate 32) has proven to be among the most invasive seaweeds worldwide. Within Australia it was first recorded from Tasmanian waters in 1988, appearing in Victoria a decade later. For the next decade it remained relatively static, probably due to some strenuous attempts to prevent its further spread, but it then appeared a few years later at Apollo Bay, from where it will have free access to Bass Strait. This is the closest port to where I live in the Otways, so it was convenient for photography, but the rate at which wakame took over the harbour is disturbing.

A winter-growing, annual seaweed, Japanese kelp produces masses of spores from the convoluted reproductive structures (sporophylls) at its base, and these attach themselves to where they will later grow within as little as 24 hours. A few weeks later the tiny attached plants, either male or female, produce drifting sperm and eggs that mate to form fertile but dormant material that won't grow until cooler conditions set in towards autumn.

Although young plants of Japanese kelp aren't hard to remove before the sporophylls appear, it is almost impossible to find them all and it doesn't take many survivors to keep the cycle going. Commercial harvesting has been mooted as a control method (retail prices for the dried product were well over $100 per kilogram at the time of writing), but as a control method this is a guaranteed failure. Apart from the central problem that any industry based on this plant would be resistant to any attempts to exterminate it, in Europe there seems little doubt that farming of this species has contributed to its further spread.

Caulerpa taxifolia (see plate 32) is a more complex problem, a 'killer algae' that has become the most successful invasive weed of the Mediterranean Sea over the past few decades, and has also invaded California. Meinesz's summary of the origins and dramatic impacts of this clone – which almost certainly originated in Australian waters – doesn't need elaboration here, and his more recent papers confirm its continued and possibly unstoppable progress (Meinesz 2001). Some sources seem to regard the Mediterranean clone as having been somehow genetically modified by aquarists (which would be a remarkable first for a seaweed), but it is most likely that this is just a vigorous strain of the species that grows particularly well in aquaria.

It is fitting to end this book with an account of this plant, a green seaweed with serrate-edged fronds (like several other indigenous species of this group), which is indisputably native in many warmer parts of Australia, and is at the same time one of the world's worst marine weeds. The so-called Mediterranean strain of this plant is an all-male clone originally dumped from an aquarium into the sea in Monaco, where there are no indigenous seaweeds of this group, and also none of the herbivorous organisms that may feed upon them in its native environments. Its hyperbolic rate of spread has raised red flags in many other countries, so that it is now illegal to grow or possess this species in southern Australia.

New populations of *C. taxifolia* have appeared around the central New South Wales coast over the past two decades, and have often been attributed to aquarium releases, although some at least are genetically closer to the natural populations in Moreton Bay than to the Mediterranean clone (see for example Schaffelke *et al.* 2002). This shouldn't be a surprise: clones collected and grown by Australian aquarists are likely to be more diverse than the Mediterranean form, simply because Australian aquarists have access to many more wild populations, but the moderate diversity in New South Wales has also been attributed to boat traffic along the east coast.

The primary reason these plants are regarded as introduced in New South Wales is because the 600 km distance south to Port Macquarie 'is considered too great a distance for the natural transportation of viable fragments' (Creese *et al.* 2004). There is lot riding on this unproven assumption, as the same report also suggests on p. 78: 'Importantly, it is not yet known whether *Caulerpa taxifolia* is having negative impacts on estuarine ecosystems and thus it is unclear whether *vast* [my italics] resources should be allocated to the control of the species.'

I would go a step further and suggest that the spread of this species is not necessarily entirely unnatural, when considered in the context of the genus *Caulerpa* as a whole. Worldwide, these plants are primarily tropical, but Australia is unique in the considerable diversity of species growing in temperate, southern waters. How they got there is no mystery over geological time periods, because the two major currents flowing southwards along our eastern and western coastlines carry numerous tropical plants and animals with them, and even the delicate yet perfectly healthy young of diverse tropical fishes are regularly carried as far south as Merimbula, almost to the Victorian border – and keep growing there. This is an ideal scenario for the gradual adaptation of subtropical species to cooler waters.

As a result of this exceptional diversity of *Caulerpa* species, I have been able to grow around a dozen temperate-zone species in aquaria over the past four decades, and can report that the serrated species are both the most adaptable, and also the most tolerant of low light levels. Floating pieces of these plants don't just survive in a poorly lit and maintained marine aquarium, they thrive and spread, suggesting that a mere matter of a few weeks drifting in a warm, well-defined and predictable southerly current is unlikely to be beyond their abilities. The most southerly indigenous forms of *C. taxifolia* are also pre-adapted to cooler temperatures, as they occur naturally in waters that may drop to around 12°C in winter.

It is unlikely that all of the weedy populations (and there are no other sorts) of *C. taxifolia* arrived spontaneously in more southerly Australian waters, especially as most of these have only been detected in the past two decades, but there may be other factors in play. Apart from anything else we need to consider that this migration may also have been helped along by the early warning effects of global warming, correlating roughly in time with other warm-water peaks that have triggered increasingly severe coral-bleaching events further north.

We will probably never know all the reasons why *C. taxifolia* is expanding its range southwards, though dumping of wild-collected clones (not just the Mediterranean form) is the most likely explanation for the South Australian populations, as it is difficult to see how these could have travelled thousands of kilometres otherwise. Along the east coast, however, it is clear that the expansion in range could at least partly be the result of the intrinsic characteristics of the plant itself, and partly another consequence of global warming. We can't afford any more aquatic weeds in this increasingly complex and turbulent world, but we also can't afford to demonise indigenous plants that are on the move, without a deeper understanding of the natural processes that may drive their spread into new places.

Glossary

allelochemicals: diverse toxic chemicals produced by some (perhaps many) plants to discourage competitors of any kind.

allelopathy: the use of diverse toxic chemicals usually by one plant against another, but also against animals including species feeding on that plant.

anaerobic: conditions in which oxygen is effectively not present, in wetland situations often creating unpleasant odours from the activities of bacteria.

annual: a plant that completes its entire lifecycle including setting seed within a year, or often less.

anoxic: low in oxygen, by contrast with anaerobic conditions where it is not present at all.

biocontrols: insects, viruses, bacteria or any other organisms that are (usually) deliberately introduced in an attempt to control weeds or other introduced species. Biocontrol must be approached with care and often requires many years of advance research, as primitive, early efforts along these lines sometimes created spectacular and unmanageable problems such as the cane toad. Also known as biological control agents.

blue-green algae: see cyanobacteria.

clone: a single, genetically identical plant, in some case of only a single sex so that it is incapable of setting seed and can only multiply vegetatively.

cosmopolitan: found naturally around most of the world.

cyanobacteria: minute plant-like bacteria that are generally better known as blue-green algae. Many species are markedly allelopathic, producing a wide range of toxins that affect not just the plants and animals sharing the same waters, but can even be dangerous to any animal that drinks from them.

disjunction: a considerable gap between different parts of the range of a plant.

drawdown: the lowering of water levels as a control method in its own right, or for easier access to submerged plants in need of control.

emergent: growing so that the foliage is mostly above water.

ephemeral: in the case of wetlands, usually drying out for part of the year so surface waters disappear.

Eurasian: found broadly from parts of Europe into western Asia, and possibly beyond.

eutrophic: nutrient-rich, often discoloured water that may be a soupy green from a rich bloom of micro-organisms.

evapo-transpiration: see transpiration.

genus (plural: genera): a formally defined group of closely related species within a plant family; there may be just one species in each genus, or a hundred.

germination: when a dormant seed starts into growth, i.e. germinates.

growing season: in the sense used in this book, a period of time during which a seasonally active plant is growing actively, and also setting seed. In southern Australia, this is usually spring to summer, sometimes extending into autumn; towards the tropics, it is more usually from when the first rains arrive, through the first few months of the dry season as wetlands dry out.

herbivore: an animal (whether fish, fowl or bug) that mainly feeds on plants.

irids: plants of the iris family.

lanceolate leaf: a relatively long and narrow leaf, with a pointed tip, in the shape of a spearhead.

leaf sheath: the lower part of each grass leaf, which clings to the stem.

monospecific: forming stands of just one species where nothing else seems able to grow.

mulch: a cover of organic matter protecting the surface of exposed ground with a material that discourages seed or plant growth.

organism: any living entity including plants, animals, fungi and bacteria.

pantropical: found through most tropical regions worldwide, including both the Old and New Worlds.

pH: the measure of acidity or alkalinity of water, starting at the neutral mark of 7, and becoming more acid as the numbers decrease from here, or more alkaline as the numbers rise. See Chapter 3 for a more detailed explanation of this term, and its applicability to weed control.

perennial: living for at least several years, in contrast to annual and biennial plants.

phytoplankton: the submerged, drifting algae and other microscopic, photosynthetic life forms that feed zooplankton, and ultimately fishes and many other aquatic animals.

rhizome: an underground stem that may run freely under water in the case of aquatics; also sometimes called a runner.

riparian: streamside; growing along the fringes of flowing waters.

slashing: mowing with a spinning bar that usually cuts higher than mowers, with a rougher finish.

substrate: the soil, gravels or other medium in which plants anchor themselves.

swale: a shallow, angled cut following a natural contour line, where water tends to soak in rather than run through, allowing nutrients to be absorbed before water enters nearby wetlands.

tepals: the petals and petal-like sepals surrounding a flower, in this book only used in the context of *Juncus* species.

transpiration: the loss of water through the leaves of plants, which can increase water losses due to evaporation, particularly in the case of such water-hungry species as willows and water hyancinth.

tuber: a swollen part of a stem or rhizome, from which new plants can grow.

turion: a dormant winter bud, which often settles to the bottom before starting into growth as temperatures rise.

vector: a carrier or spreader of undesirable plants, their seed, or of disease.

vegetative reproduction: production of genetically identical plants from runners, proliferations, and other offsets but not by seed.

viable: when referring to seed, the able to germinate or grow. Viability is often expressed as a percentage, for example 20 per cent meaning only one seed in five is likely to grow, and these percentages usually drop as the seed ages.

whorl: a circle of leaves surrounding a central stem, with the same pattern repeating at regular intervals to form a tidy plume sometimes referred to as a foxtail.

zooplankton: mostly minute, drifting animals that are a significant food source for many other aquatic animals.

Further information including websites

To avoid unnecessary duplication, only relatively recent publications dealing with various wetland weeds are included, as these already contain most of the older literature in their references. I have also listed some additional references that are not as widely known. Some exceptions have been made, including species-specific articles from two volumes of *The Biology of Australian Weeds*, which were a useful and fairly current summary of what was known of about a number of significant weeds at the time of publication – and are now a salutary warning as to just how fast some weeds can spread in less than 15 years. Only the original publication date or edition of any book cited here is given, unless a more recent edition with *new* information specifically on wetlands has been published since that time.

Websites have been a more complex problem as they vary dramatically in their usefulness, and entries often disappear, or at the very least can become difficult to find. The major sites include the National Environmental Alert List at <www.weeds.gov.au/weeds/lists/alert.html>, with National Sleeper Weeds at <www.weeds.gov.au/weeds/lists/sleeper.html>. These sites include many specific weed management guides and are also often useful for their links. The Weeds of National Significance site is at <www.weeds.org.au/natsig.htm>. However, even on these reputable sites the information is not always consistent, or even necessarily always accurate, as will be noted from some entries in the compendium of weeds.

References

Adair R, Cheal D and White M (2008) 'Advisory list of environmental weeds of aquatic habitats in Victoria'. Department of Sustainability and Environment, Melbourne.

Anon (2009) 'New South Wales control plan for the noxious marine alga *Caulerpa taxifolia*'. Department of Industry and Investment, Orange, NSW.

Anon (2009) *Blackberry Control Manual: Management and Control Options for Blackberry* (Rubus *spp.*) *in Australia*. New South Wales Department of Primary Industries Weed Management Unit, Orange, NSW.

Anon (Eds) (2009) *Plants Behaving Badly: In Agriculture and the Environment. Proceedings Fourth Victorian Weed Conference*. Frankston, (Melbourne) Victoria.

Aston HI (1973) *Aquatic Plants of Australia*. Melbourne University Press, Melbourne.

Barker RD and Vestjens WJM (1989) *The Food of Australian Birds* (2 volumes). CSIRO Publishing, Melbourne.

Blood K (2001) *Environmental Weeds: A Field Guide for South-Eastern Australia*. Science Publishers, Melbourne.

Brock MA, Boon PI and Grant A (Eds) (1994) *Plants and Processes in Wetlands*. CSIRO Australia, Canberra.

Chambers JM, Fletcher NL and McComb AJ (1995) *A Guide to Emergent Wetland Plants of South-western Australia*. Murdoch University, Perth.

Cook CDK (1990) *Aquatic Plant Book*. SPB Academic Publishing, The Netherlands.

Cowie ID, Short PS and Osterkamp Madsen M (2000) *Floodplain Flora: A Flora of the Coastal Floodplains of the Northern Territory, Australia*. Australian Biological Resources Study, Canberra.

Creese RG, Davis AR and Glasby TM (2004) 'Eradicating and preventing the spread of the invasive alga *Caulerpa taxifolia* in New South Wales'. New South Wales Fisheries Report No. 35593, Final Report Series No. 64, Cronulla.

Cremer K, Van Kraayenoord C, Parker N and Streatfield S (1995) Willows spreading by seed – implications for Australian river management. *Australian Journal of Soil and Water Conservation* **8**(4), 18–27.

Cronk JK and Fennessy MS (2001) *Wetland Plants: Biology and Ecology*. Lewis Publishers, Boca Raton, Florida.

Csurhes S (2008) 'Pest plant risk assessment: Glush Weed, *Hygrophila costata*'. Department of Primary Industries, Brisbane.

D'Costa D and Kershaw AP (1997) An expanded recent pollen database from south-eastern Australia and its potential for refinement of paleoclimatic estimates. *Australian Journal of Botany* **45**, 583–605.

Dodson JR (1979) Late Pleistocene vegetation and environments near Lake Bullenmerri, western Victoria. *Australian Journal of Ecology* **4**, 419–427.

Dodson JR (1983) Modern pollen rain in south-eastern New South Wales, Australia. *Review of Palaeobotany and Palynology* **38**, 249–268.

Dodson JR (1986) Holocene vegetation and environments near Goulburn, New South Wales. *Australian Journal of Botany* **34**, 231–249.

Dodson JR and Wilson IB (1975) Past and present vegetation of Marshes Swamp in south-eastern South Australia. *Australian Journal of Botany* **23**, 123–150.

Entwisle TJ, Sonneman JA and Lewis SH (1997) *Freshwater Algae in Australia: A Guide to Conspicuous Genera.* Sainty & Associates, Sydney.

Finlayson CM, Roberts J, Chick AJ and Sale PJM (1995) *Typha domingensis* Pers. and *T. orientalis* Presl. In: *The Biology of Australian Weeds Volume 1.* (Eds RH Groves, RCH Shepherd and RG Richardson) pp. 231–242. RG & FJ Richardson, Melbourne.

Given DR (1994) *Principles and Practice of Plant Conservation.* Timber Press, Oregon.

Haaksma ED and Linder PL (2000) *Restios of the Fynbos.* The Botanical Society of South Africa, Cape Town.

Halliday I (2004) *Melaleucas: A Field and Garden Guide.* New Holland Publishers, Sydney.

Harden GJ (Ed.) (1990–1993) *Flora of New South Wales* (volumes 1–4). New South Wales University Press, Sydney.

Hibbert M (2004) *The Aussie Plant Finder.* Florilegium, Sydney.

Hiscock P (2003) *Encyclopedia of Aquarium Plants.* Barron's Educational Series, New York.

Hocking PJ, Finlayson CM and Chick AJ (1998) *Phragmites australis* (Cav.) Trin. ex Steud. In: *The Biology of Australian Weeds Volume 2.* (Eds FD Panetta, RH Groves and RCH Shepherd) pp. 177–197. RG & FJ Richardson, Melbourne.

Holderread D (2001) *Storey's Guide to Raising Ducks.* Story Books, Vermont.

Holland-Clift S and Davies J (2007) *Willows National Management Guide: Current Management and Control Options for Willows* (Salix *spp.*) *in Australia.* Victorian Department of Primary Industries, Geelong.

Huisman JM (2000) *Marine Plants of Australia.* University of Western Australia Press, Perth.

Hussey BMJ, Keighery GJ, Dodd J, Lloyd SG and Cousens RD (2007) *Western Weeds: A Guide to the Weeds of Western Australia.* 2nd edn. The Plant Protection Society of Western Australia, Perth.

Johnson PN and Brooke PA (1989) *Wetland Plants in New Zealand.* DSIR Publishing, Wellington.

Julien MH (1995) *Alternanthera philoxeroides* (Mart.). In: *The Biology of Australian Weeds Volume 1.* (Eds RH Groves, RCH Shepherd and RG Richardson) pp. 1–12. RG & FJ Richardson, Melbourne.

Kasselmann C (2003) *Aquarium Plants.* Krieger Publishing Company, Florida.

Kloot PM (1984) The introduced elements of the flora of southern Australia. *Journal of Biogeography* **11**, 63–78.

Lamp CA, Forbes SJ and Cade JW (2001) *Grasses of Temperate Australia: A Field Guide.* 2nd edn. Bloomings Books, Melbourne.

Lazarides M, Cowley K and Hohnen P (1997) *CSIRO Handbook of Australian Weeds*. Australian National Herbarium Centre for Plant Biodiversity Research, Canberra.

Lewis JA (1999) A review of the occurrence of exotic macroalgae in southern Australia, with emphasis on Port Phillip Bay, Victoria. In: *Marine Biological Invasions of Port Phillip Bay, Victoria*. (Eds CL Hewitt, ML Campbell, RE Thresher and RB Martin) pp. 61–86. CSIRO Marine Research, Hobart.

Lonsdale WM, Miller IL and Forno IW (1995) *Mimosa pigra* L. In: *The Biology of Australian Weeds Volume 1*. (Eds RH Groves, RCH Shepherd and RG Richardson) pp. 169–188. RG & FJ Richardson, Melbourne.

Lumpkin TA and Plucknett DL (1980) *Azolla*: botany, physiology and use as a green manure. *Economic Botany* **34**(2), 111–153.

Mackey AP and Swarbrick JT (1998) *Cabomba carolineana* A. Gray. In: *The Biology of Australian Weeds Volume 2*. (Eds FD Panetta, RH Groves and RCH Shepherd) pp. 19–35. RG & FJ Richardson, Melbourne.

McDowall RM (1989) *New Zealand Freshwater Fishes: A Natural History and Guide*. 2nd edn. Heinemann Reed Publishing Group, Auckland.

Meinesz A (2001) *Killer Algae: The True Tale of a Biological Invasion*. The University of Chicago Press, Chicago.

Muehlberg H (1981) *The Complete Guide to Water Plants*. EP Publishing, London.

Nash H and Stroupe S (1998) *Aquatic Plants and Their Cultivation: A Complete Guide for Water Gardeners*. Sterling Publishers, New York.

Parsons WT and Cuthbertson EG (2001) *Noxious Weeds of Australia*. 2nd edn. CSIRO Publishing, Melbourne.

Pieterse AH and Murphy KJ (1990) *Aquatic Weeds: The Ecology and Management of Nuisance Aquatic Vegetation*. Oxford University Press, Oxford.

Randall RP (2007) *The Introduced Flora of Australia and its Weed Status*. CRC for Australian Weed Management, Adelaide.

Rataj K and Horeman TJ (1977) *Aquarium Plants: Their Identification, Cultivation and Ecology*. Tropical Fish Hobbyist Publications, New Jersey.

Richardson FJ, Richardson RG and Shepherd RCH (2006) *Weeds of the South-East: An Identification Guide for Australia*. RG & FJ Richardson, Meredith, Victoria.

Rizvi SJH and Rizvi V (1992) *Allelopathy: Basic and Applied Aspects*. Chapman and Hall, London.

Romanowski N (1992) *Water and Wetland Plants for Southern Australia*. Lothian Books, Melbourne.

Romanowski N (1998) *Aquatic and Wetland Plants: A Field Guide for Non-Tropical Australia*. University of New South Wales Press, Sydney.

Romanowski N (2000) *Water Garden Plants and Animals: The Complete Guide for All Australia*. University of New South Wales Press, Sydney.

Romanowski N (2007) *Edible Water Gardens: Growing Water-Plants for Food and Profit*. Hyland House, Melbourne.

Romanowski N (2009) *Planting Wetlands and Dams: A Practical Guide to Wetland Design, Construction and Propagation.* 2nd edn. Landlinks Press, Melbourne.

Room PM and Julien MH (1995) *Salvinia molesta* D.S. Mitchell. In: *The Biology of Australian Weeds Volume 1.* (Eds RH Groves, RCH Shepherd and RG Richardson) pp. 217–230. RG & FJ Richardson, Melbourne.

Sainty GR and Jacobs SWL (2003) *Waterplants in Australia: A Field Guide.* 4th edn. Sainty and Associates, Sydney.

Schaffelke B, Murphy N and Uthicke S (2002) Using genetic techniques to investigate the sources of the invasive alga *Caulerpa taxifolia* in three new locations in Australia. *Marine Pollution Bulletin* **44**(3), 204–210.

Sindell BM (Ed.) (2000) *Australian Weed Management Systems.* RG & FJ Richardson, Melbourne.

Slocum PD and Robinson P (1996) *Water Gardening: Water Lilies and Lotuses.* Timber Press, Portland, Oregon.

Smith WH and Cervantes EP (Eds) (1987) *Azolla Utilization.* International Rice Research Institute, Los Banos, Philippines.

Soerjani M, Kostermans AJGH and Tjitrosoepomo G (1987) *Weeds of Rice in Indonesia.* Balai Pustaka, Jakarta.

Stock D (Ed.) (2009) *Lantana: Best Practice Manual and Decision Support Tool.* Department of Employment, Economic Development and Innovation, Queensland.

Streever WJ (Ed.) (1997) Wetland rehabilitation in Australia. *Wetlands Ecology and Management* special issue **5**(1), 1–97.

Swarbrick JT, Finlayson CM and Cauldwell AJ (1995) *Hydrilla verticillata* (L.f.) Royle. In: *The Biology of Australian Weeds Volume 1.* (Eds RH Groves, RCH Shepherd and RG Richardson). RG & FJ Richardson, Melbourne.

Swarbrick JT, Wilson BW & Hannan-Jones MA (1998) *Lantana camara* L. In: *The Biology of Australian Weeds Volume 2.* (Eds FD Panetta, RH Groves & RCH Shepherd) pp. 82–95. RG & FJ Richardson, Melbourne.

van Oosterhout E (2007) *Alligator Weed Control Manual: Eradication and Suppression of Alligator Weed (*Alternathera philoxeroides*) in Australia.* NSW Department of Primary Industries, Sydney.

Walsh NG and Entwistle TJ (Eds) (1996–1999) *Flora of Victoria* (volumes 2–4). Inkata Press, Melbourne.

Walsh NG and Stajsic V (2007) *A Census of the Vascular Plants of Victoria.* 8th edn. Royal Botanic Gardens, Melbourne.

Wright AD and Purcell MF (1995) *Eichhornia crassipes* (Mart.) Solms-Laubach. In: *The Biology of Australian Weeds Volume 1.* (Eds RH Groves, RCH Shepherd and RG Richardson) pp. 111–121. RG & FJ Richardson, Melbourne.

Index

www.ingramcontent.com/pod-product-compliance
Lightning Source LLC
LaVergne TN
LVHW060630110826
845147LV00014B/887

9780643103955